Lady Eugenist

BOOKS FROM INKLING

Free Lover: Sex, Marriage and Eugenics in the Early Writings of Victoria Woodhull by Victoria Woodhull with introductions by Michael W. Perry. Contains her early speeches critical of marriage and in support of free love and a mystical eugenics. Includes: "And the Truth shall make you Free: A Speech on The Principles of Social Freedom," "The Scarecrows of Sexual Slavery, An Oration," "The Elixir of Life or, Why Do We Die?, An Oration," and "Tried as by Fire or, The True and False, Socially."

ISBN: 1-58742-050-3 (pb), 1-58742-051-1 (hb) & 1-58742-052-x (ebook)

The Pivot of Civilization in Historical Perspective by Margaret Sanger with articles by Victoria Woodhull, H. G. Wells, G. B. Shaw, Theodore Roosevelt, Ellen Key, Henry Goddard, G. K. Chesterton, Charlotte Perkins Gilman, Archbishop Patrick Hayes, and Oliver Wendell Holmes, Jr. Includes the entire text of Margaret Sanger's 1922 *The Pivot of Civilization* plus 31 chapters of prologue containing numerous original source documents from the era. To understand the early birth control movement in the United States, this book is a must-have. (Also available as ebooks.)

ISBN: 1-58742-004-x (paperback) & 1-58742-008-2 (hardback)

Eugenics and Other Evils by G. K. Chesterton. In the early twentieth century eugenics was considered enlightened, scientific, and progressive. Chesterton wrote one of the few book-length criticisms of eugenics published during that era. This edition includes appendices by eugenists such as Francis Galton, C. W. Saleeby, and Dr. Marie Stopes.

ISBN: 1-58742-002-3 (paperback) & 1-58742-006-6 (hardback)

Victoria Woodhull ebooks: *Children—Their Rights and Privileges* (1871) with selections from "Press Notices" in *The Human Body the Temple of God* (1890), ISBN: 1-58742-043-0. *The Garden of Eden* (1875/1890), ISBN: 1-58742-044-9. *Stirpiculture* (1888), ISBN: 1-58742-045-7. *Humanitarian Government* (1890), ISBN: 1-58742-046-5. *The Rapid Multiplication of the Unfit* (1891), ISBN: 1-58742-047-3. *The Scientific Propagation of the Human Race* (1893), ISBN: 1-58742-048-1. *A Fragmentary Record of Public Work Done in America, 1871–1877* (1887), ISBN: 1-58742-049-x.

To order online visit: http://www.InklingBooks.com/

Available internationally through Amazon.com and other online stores.

Bookstores can order these books from Baker & Taylor or Ingram in the U.S.A. or Bertrams in Europe.

Lady Eugenist

Feminist Eugenics
in the Speeches and Writings
of Victoria Woodhull

with

Children—Their Rights and Privileges

The Human Body (selected "Press Notices")

The Garden of Eden

Stirpiculture

Humanitarian Government

The Rapid Multiplication of the Unfit

The Scientific Propagation of the Human Race

by

Victoria Woodhull

With Introductions by Michael W. Perry

Inkling Books   Seattle   2005

Description

This book contains facsimile reproductions of original editions of five (31–68 page) published speeches on eugenics by Victoria C. Woodhull (Chapters 4–8), a newly typeset version of a published speech too darkened to reproduce (Chapter 2), and a selection of newspaper articles and letters about Victoria Woodhull as published by her (Chapter 3). The books are arranged in chronological order by their original publication date. There is also an introductory chapter and introductions to each Woodhull speech by Michael W. Perry. Although not listed in many dictionaries today, the spelling "eugenist" used here was common in the early twentieth century, as the back cover graphic illustrates.

Copyright Notice

© Copyright 2005, Michael W. Perry, All Rights Reserved. No part of this book may be reproduced, stored in a retrieval system, or transmitted in any form or by any means (electronic, mechanical, photocopy, recording or otherwise) without the express written permission of the copyright holder with the sole exception of fair use provisions recognized by law.

For Teachers: For permission to reproduce more than 20 pages of this book, contact the publisher through the Internet address below. In addition to paperback and hardback editions, an inexpensive, printable e-book is available for students from online sites such as Amazon.com. Individual books in this collection are also available as separate e-books.

Publisher's Note

"Children—Their Rights and Privileges" (Chapter 2) is based on a speech given to the American Association of Spiritualists at its Eighth National Convention on Wednesday, September 13, 1871 in Troy, NY. This text comes from the October 7, 1871 issue of *Woodhull & Claflin's Weekly* and was originally entitled "The Training of Children—Good Advice to Mothers."

"The Garden of Eden" (Chapter 4) was first published in 1875. This version is from *The Human Body The Temple of God* published in 1890 London, as were the selections from newspapers and letters included in Chapter 3, "Press Notices."

Stirpiculture or, The Scientific Propagation of the Human Race (Chapter 5) was published in London in February of 1888. Includes "Some Thoughts About America"

Humanitarian Government (Chapter 6) was published in 1890 London.

The Rapid Multiplication of the Unfit (Chapter 7) was published in 1891 London.

The Scientific Propagation of the Human Race; or, Humanitarian Aspects of Finance and Marriage. The Science of Well Being (Chapter 8) was published in 1893 and contained no place of publication. University library databases suggest both New York and the Eugenics Society in London. Includes "Efficacy of Punishments" (originally published in *The Humanitarian*, August 1892).

Special Thanks

Special Collections at Southern Illinois University, Harvard University, Boston University, Yale University, University of Washington, New York Public Library, Library of Congress, King County (Washington) Library System, Seattle Public Library, and Mary Shearer of VictoriaWoodhull.com.

Library Cataloging Data

Lady Eugenist: Feminist Eugenics in the Speeches and Writings of Victoria C. Woodhull

Victoria C. [Claflin] Woodhull (1837–1927) with introductions by Michael W. [Wiley] Perry (1948–)

Other published names for the author include Victoria Woodhull Martin and Mrs. John Biddulph Martin

Includes all the text of the previously published speeches as listed above.

331 pages, Size: 6 x 9 inches or 229 x 152 mm Thickness: 0.75 inches or 19 mm (paperback) and 0.88 inches or 23 mm (hardback)

Library of Congress Control Number: 2005929663

ISBN-10: 1-58742-040-6   ISBN-13: 978-1-58742-040-5 (alkaline paperback)

ISBN-10: 1-58742-041-4   ISBN-13: 978-1-58742-041-2 (alkaline hardback)

ISBN-10: 1-58742-042-2   ISBN-13: 978-1-58742-042-9 (Adobe PDF ebook)

BISAC Subject Headings: HIS054000 HISTORY/Social History; HIS036040 HISTORY/United States/19th Century; SCI075000 SCIENCE/Philosophy & Social Aspects

Publisher Information

First Inkling Books edition, Second Printing, November, 2005. Published in the United States of America on acid-free paper. Inkling Books, Seattle, WA, U.S.A. Internet: http://www.InklingBooks.com/

Contents

1. Was Victoria Woodhull the First Eugenist?.......... 7
2. Children—Their Rights and Privileges (1871)
    Introduction 27
    Text...30
3. Press Notices (1869–1882, publ. 1890).............. 43
4. The Garden of Eden (1875, publ. 1890)
    Introduction 57
    Text...63
5. Stirpiculture (1888)
    Introduction 119
    Text..121
    "Some Thoughts about America"....................143
6. Humanitarian Government (1890)
    Introduction 151
    Text... 156
7. The Rapid Multiplication of the Unfit (1891)
    Introduction 223
    Text... 231
8. The Scientific Propagation of the Human Race (1893)
    Introduction 269
    Text... 273
    "Efficacy of Punishments" (1892)..................314

Chapter 1

Was Victoria Woodhull the First Eugenist?

by Michael W. Perry

In both Europe and the United States, the nineteenth century was an exceptionally fertile time for ideas. For good and ill, all-encompassing schemes for social reorganization, such as nationalism and socialism, were nurtured until they grew strong—sadly, often strong enough to be dangerous, as the twentieth century would demonstrate.

One of those ideas was eugenics. The basic idea of eugenics, controlled breeding, has no inventor. Long before written history, people domesticated animals, selecting the most useful to parent the next generation. Cattle were chosen for meat or milk, chickens for eggs, horses for strength, and dogs for their eagerness to obey.

In historical times and particularly under the influence of Christianity, however, there was an unwillingness to extend breeding to people. For all their power, Europe's feudal lords could not mate their strongest peasant lad with their sturdiest lass, nor could they forbid the weakest from marrying and parenthood. People, however poor, still bore the image of God, and marriage was a holy sacrament. Eugenists would even complain that Medieval Catholicism had been anti-eugenic, drawing the most talented into lives of celibacy.

Other societies were different. Plato's utopia, described in his *Republic*, was to be eugenic. Nearby Sparta actually practiced a ruthless form of eugenics, killing infants who seemed unlikely to become hardy warriors. Sparta demonstrates one problem with applying ideas that work with farm animals to people. With a chicken, you know what you want—eggs. But what do you want a human to be? Isaac Newton, born tiny and premature, would have been quickly discarded in Sparta, as would Theodore Roosevelt and Winston Churchill, both sickly children who grew up to become strong and talented leaders.

But there is a sense in which eugenics has a pioneer. Pioneering an idea is like exploration and discovery. In the strictest sense, the person who discovered America was the unknown person who first crossed the land bridge from Asia. Those who want to add to "discover America" an implicit "from Europe," get caught up in debates about dates and evidences. Where did the Norse settle? Did Irish monks come here? Did Phoenicians arrive before anyone else? Columbus is unique only in that he made his exploration well known. If his three ships had been forced to return, having found no land to the west, his reputation would have been ruined. So, in that sense and even though what he found wasn't what he was looking for, Columbus discovered America. He staked his reputation on land being there and was proved right openly and publicly. That's why the world was never the same after him. It wasn't changed by Norse or Irish explorers.

That's the sense in which this book suggests Victoria Woodhull pioneered eugenics, and it's the same sense she herself claimed. She wasn't the first to come up with the idea, or the first the first to write about it. She did not get a few people to live eugenically in the small Oneida community. But she may have been the first to stake her reputation on eugenics becoming a cause. In that sense, she is the "Lady Eugenist" of this book's title, and everyone who came after, whether they admitted it or not, were her followers.

Unfortunately, that's not the official line. Almost without exception, the story of eugenics as told by historians has favored well-born, well-to-do, well-educated men, mostly English. Perhaps deceived by eugenists, who found Woodhull's prior claim embarrassing, historians have neglected the role played by women, particularly American women such as Victoria Woodhull, Charlotte Perkins Gilman, and Margaret Sanger. While I disagree strongly with virtually every premise and practice of eugenics, I believe this silence is unfair. Woodhull wanted to be credited as a eugenic pioneer—perhaps even *the* eugenic pioneer—and, as this book will demonstrate, she earned that title.

Background

First, I should explain how this book came to be. In early 2001, I was researching my second book on eugenics, *The Pivot of Civilization in Historical Perspective*[1] (from here on referred to as *Pivot*). With it, I intended to do something few had done. I wanted to take Planned Parenthood's founder, Margaret Sanger, seriously as a thinker rather than just a controversial activist. I wanted to show how what she believed fit into a fierce debate about differing birthrates in early twentieth-century America. In short, I wanted to give her a mind.

One chapter in that book[2] looked briefly (four pages) at what Victoria Woodhull had written on eugenics in an 1891 booklet, *The Rapid Multiplication of the Unfit*, republished here in its entirety (Chapter 7). At the time, that did not strike me as particularly important. The year 1891 is, after all, relatively late in the accepted history of modern eugenics. But it was while researching a different chapter that I came across something that startled me.

Chapter XII in *Pivot*, "How Bright a Torch," describes the open enthusiasm the *New York Times* once displayed for eugenics, particularly in 1912 when its coverage of the topic peaked, perhaps because of a prestigious eugenics conference held in London that summer. Given the newspaper's enthusiasm for getting rid of those it regarded as unfit, I decided to see how it covered *Buck*

---

1. Michael W. Perry, ed. *The Pivot of Civilization in Historical Perspective* (Seattle: Inkling Books, 2001). Page references to *Pivot* are to the wide-format paperback rather than the hardback.
2. *Pivot*, Chapter IV, "The Rapid Multiplication of the Unfit."

v. Bell, the infamous 1927 U.S. Supreme Court decision that declared eugenic sterilization constitutional. Had its zeal waned in the intervening fifteen years?

It had not. Speaking as if it defined all that was true and proper, the newspaper could not have been more approving. Two of the headlines were "Right to Protect Society" and "Justice Holmes Draws Analogy to Compulsory Vaccination in Woman's Case." The *Times* clearly agreed with the author of the opinion. Had it thought otherwise, it could have pointed out a critical difference between a Massachusetts vaccination law cited in the opinion and Virginia's sterilization law. Anyone opposed to being vaccinated could pay a five dollar fine and be left alone. The sterilization law offered no such escape. And so no one would think that forced eugenic sterilization was in the least controversial, the *Times* buried the story deep inside the paper near an article about another court's decision to ban the use of "chain coupons" in "silk stockings sales."³

The paper's coverage did not end there. Five days later, the *Times* ran an article by an Associated Press reporter in Brighton, England who, gushing with praise, had interviewed Victorian Woodhull, now bearing the name Martin. What drew my attention were the following remarks. (Bolding added.)

… Time has not dimmed the eyes of this spirited woman who, with her sister, the late Lady Cook, formerly Tennessee Claflin, was the first woman broker in New York and lectured and published *Claflin's Weekly* **in support of equal suffrage and eugenics before they both came to England**.…

Mrs. Martin, who **wrote and lectured for thirty years on eugenics,** remarked that she was pleased to read that the Virginia Eugenics law had succeeded in establishing the right to sterilize the feeble-minded.

"**I advocated that fifty years ago** in my book, *Marriage of the Unfit*," she said. "I am also glad that parents are now beginning to instruct their adolescent children in the facts of life. My sister, Tennessee, and I were mercilessly slandered fifty years ago, when **we dared to advocate women's emancipation and discussed eugenics** in America, but time has proved that we were right."⁴

I was surprised by two things Woodhull said. The first was her claim to have been promoting eugenics some "fifty years ago" (1877), which she said was before it "came to England." That suggested she promoted it in the U.S. through her public lectures and in *Woodhull and Claflin's Weekly* (1870–76), and that she had brought those ideas with her when she moved to England in 1877.

I wondered at the time if she was being honest or just playing her usual "I was the first woman to…" game. I have since discovered she was right. A speech

---

3. "Upholds Operating on Feeble-Minded." *New York Times* (May 3, 1927), 19. In *Pivot*, 31.
4. "Says Voting at 25 is 'Young Enough.'" *New York Times* (May 3, 1927), pt. 2, p. 6. In *Pivot*, 31. For a facsimile, see page 56 in this book. I could not find *Marriage of the Unfit*. She may have meant *The Rapid Multiplication of the Unfit*, or it may have been a pamphlet printed in the 1870s in such small quantities that no copy survives today.

she made in 1871, republished in Chapter 2 of this book, is clearly eugenic. Newspapers from the 1870s, quoted in Chapter 3, mention eugenic ideas in speeches she gave across the U.S. Finally, as you can see from the title page of last booklet in this collection (Chapter 8), as least as early 1893 she claimed to have lectured on eugenics "throughout America, from 1870 to 1876."

The second odd remark was that Woodhull and her sister "dared to advocate women's emancipation and discussed eugenics." How could she link emancipation with eugenics? For eugenics to succeed, ten to twenty percent of women must be kept from having any children, and perhaps another twenty to thirty percent must be kept from having more than two. That's regimentation rather than emancipation. Later we'll see what she meant.

The Role of Francis Galton

Although Francis Galton (1822–1911) would not coin the term "eugenics" until 1883, historians tell us that eugenics began in England with Galton's 1865 article in *Macmillan's Magazine* entitled "Hereditary Talent and Character." That's why I included extracts from it as the first appendix in my edition of G. K. Chesterton's 1922 *Eugenics and Other Evils*.[5]

Woodhull's own remarks, however, suggest the history of eugenics may be more complicated. Was Galton's early article that significant, or was there an independent history of eugenics in the United States, one that helped start the movement in England, and one whose chief spokesman was a woman from a far less prestigious background than Galton? After all, the idea of breeding people like farmers mate their livestock isn't hard to imagine, however distasteful many of us now find it. It needs neither Darwin nor Galton for inspiration.

From Galton's day until the present, eugenists certainly have not been happy about suggestions that they might be linked to Woodhull, a controversial speaker-for-hire. In a 1976 book supported by the Council of the Eugenics Society, G. R. Searle dismisses Woodhull as little more than an "engaging charlatan." "Stirpiculture" (Chapter 5) and "scientific propagation" (Chapter 8) were two early terms for what later came to be called "eugenics." (Bolding added.)

> Another embarrassment to sober men like Galton was the American, Victoria Woodhull-Martin, an engaging charlatan, whose bizarre career took her through three marriages, numerous liaisons, several well-publicized scandals, an attempt to stand as President of the United States, and advocacy of spiritualism, elixirs of life, communism, sex equality, free love—**and stirpiculture.** But, despite her invocations of "science," Mrs. Woodhull-Martin had no authority to pontificate on matters of human inheritance, and many of her observations on this subject were ill-formed and nonsensical. **It was the backing of responsible and established scientific men which was essential to the progress of eugenics.** All this had to wait until

---
5. G. K. Chesterton, *Eugenics and Other Evils* (Seattle: Inkling Books, 2000), 123–26.

such a time as biologists had acquired an understanding of heredity that would enable them to explain how parents transmitted certain of the physical and intellectual qualities to their offspring. Not until the Edwardian period had the scientific groundwork been sufficiently well laid for eugenics to become a plausible political creed.[6]

"Established scientific men"—that's like saying that discoveries of new lands can only be done by those who belong to the Royal Geographic Society. Uncredentialed men and particularly women do not count, no matter how early they visit or how widely they describe their travels when they return. Yes, some of Woodhull's ideas were "bizarre," and I don't hesitate to point that out. But Columbus' geography was so bizarre, he thought he had reached the East Indies, even though an entire continent and an ocean stood in his way. And most of Woodhull's eugenic ideas weren't that out of line with what the scientists and medical men of her day believed. Some stand up to scrutiny at least as well as those championed by the more scientifically credentialed eugenists. Here's what Searle sniffed about Woodhull in an endnote.

> Many of her 'scientific' theories, in fact, ran counter to the central propositions of the eugenists, e.g. 'The most active agent in generating the unfit is fatigue poison...'; much 'family generation... is due to physical exhaustion from overwork or the lack of sufficient light and fresh air;' see V. C. Woodhull-Martin, *The Rapid Multiplication of the Unfit* (1891), p. 10. It is Mrs. Woodhull-Martin's oft-repeated contention that those who were unfit through fatigue produced degenerate offspring. Although she lived the latter part of her life in London, where she edited a weekly, *The Humanitarian*, H. G. Wells is the only British writer known to me to acknowledge any debt to her; see his *Mankind in the Making* (1903), p. 39.

Notice that Serle's main objection to Woodhull was that what she said "ran counter to the central propositions of the eugenists" that hereditary was everything and environment counted for little. Unlike Woodhull, who seemed to grab ideas from everywhere with little concern for their source, many eugenists were a prissy lot, liking their theories wrapped up in neat packages. But few in medicine today would fault her for attacking "fatigue poison." Exhaustion is an indication that a child's parents aren't getting nutritious food, enough rest, or "sufficient light and fresh air," for their unborn child's health. An unborn baby may not be influenced by a mother in all the ways Woodhull assumed. But she was right in her claim that the influence wasn't simply genetic.

Given the importance of H. G. Well's own popularization of eugenics from 1901 on, if Woodhull had only influenced him, she would have accomplished

---

6. G. R. Searle, *Eugenics and Politics in Britain 1900–1914* (Leyden: Noordhoff International Publishing, 1976), 5–6. The Edwardian period ran from 1901–10, with some extending it to World War I. Notice those "responsible and established scientific men" were waiting for evidence to render "plausible" a "political creed" they already believed—in the case of Galton since 1865. That's more ideological than scientific.

quite a bit. The year before *Mankind in the Making* (1903) was published as a book, it was serialized in *Cosmopolitan* magazine, reaching the American equivalent of what Wells called "titled ladies of liberal outlook." (Those women later provided support for Margaret Sanger's birth control movement.) But Wells credited Woodhull with more than that. He contrasted Francis Galton's 1901 Huxley Lecture to the Anthropological Institute, which was heard by a few, with Woodhull's earlier lectures and writings before a much wider audience. Of course, he also noted that the queer "absurdity and pretentiousness" of many the writers she found for her cause made him wish it would be taken up by "sober and honest men"—a term that in Wellesian meant ruthless and unprincipled men who knew how to brush aside opposition and get things done.

> At a more popular level Mrs. Victoria Woodhull Martin has battled bravely in the cause of the same foregone conclusion. The work of telling the world what it knows to be true will never want self-sacrificing workers. *The Humanitarian* was her monthly organ of propaganda. Within its cover, which presented a luminiferous stark ideal of exemplary muscularity, popular preachers, popular bishops, and popular anthropologists vied with titled ladies of liberal outlook in the service of this conception. There was much therein about the Rapid Multiplication of the Unfit, a phrase never properly explained, and I must confess that the transitory presence of this instructive little magazine in my house, month after month (it is now, unhappily, dead), did much to direct my attention to the gaps and difficulties that intervene between the general proposition and its practical application by sober and honest men. One took it up and asked time after time, "Why should there be this queer flavour of absurdity and pretentiousness about the thing?" Before the *Humanitarian* period I was entirely in agreement with the *Humanitarian*'s cause. It seemed to me then that to prevent the multiplication of people below a certain standard, and to encourage the multiplication of exceptionally superior people, was the only real and permanent way of mending the ills of the world. I think that still. In that way man has risen from the beasts, and in that way men will rise to be over-men. In those days I asked in amazement why this thing was not done, and talked the usual nonsense about the obduracy and stupidity of the world. It is only after a considerable amount of thought and inquiry that I am beginning to understand why for many generations, perhaps, nothing of the sort can possibly be done except in the most marginal and tentative manner.[7]

Wells was a eugenic moderate, skeptical that, given our current state of knowledge and our existing political institutions, much could be done to improve the world "except in the most marginal and tentative manner." He would express similar doubts in the introduction he wrote for Margaret Sanger's 1922 *The Pivot of Civilization*. Wells rested his hope in a world run by a technocratic elite that would adopt a clever but covert eugenic scheme. From 1901 on, in

---

7. H. G. Wells, *Mankind in the Making* (New York: C. Scribner's Sons, 1904), 36–7.

novel after novel, he wrestled with how that sort of government was to be established. Until that day came, he intended to soothe the great bulk of his readers with claims that, even when eugenics was applied, it would change little. In the paragraph following the one quoted above, he claimed that if Galton and Woodhull were given the power to form a committee that would dictate who could marry and have children, it would in the end "decide to leave matters almost exactly as they are now." That's nonsense. Woodhull approved schemes in which scientific committees forcibly sterilized poor women. Galton was writing, as early as his seminal 1865 article, of using "any agency" that would retard marriages in "caste B," which he described as "the refuse" of society.[8]

Woodhull in 1912 London

My interest in Woodhull's role grew when the Victoria Woodhull collection at Southern Illinois University graciously suppled me with two articles from July 29, 1912 issues of London newspapers. Both were part of the news coverage of the International Eugenics Congress then being held in the city.[9]

The first article, "Lady Eugenist," (page 16 in this book) is from *Pall Mall Gazette* and provided the *Lady Eugenist* title for this book. The two quotes are from her 1893 *The Scientific Propagation of the Human Race* (Chapter 8). Without indicating so, the first links text from pages 11 and 12 (in the original) and the second comes from page 13 with "Idealise sexual selection and" left out. The second article, "Eugenics in London Thirty-five Years Ago," (page 17) is apparently from the *Evening Standard*. Since the two share a common theme but have almost no details in common, it's possible the same cash-starved correspondent wrote both, making sure each paper got a different story.

Notice from "Lady Eugenist," that the idea that Francis Galton was "responsible for the interest we now take in Eugenics," was already well established in 1912. It certainly did the cause good to have such a distinguished scientist as a founder. But the article's author, probably encouraged by Woodhull herself, thought that unfair: "Galton can only claim credit for having, by a large sum which he left, brought the present International Eugenics Congress together."

Interestingly, Galton, who had died the year before, seemed to agree. In his 1908 autobiography, *Memories of My Life*, he opened his chapter on "Race Improvement" with revealing frankness. (*Hereditary Genius* was published in 1869 with a second edition in 1892. Bolding added.)

---
8. For more on Galton's grand scheme, see Inkling's *Eugenics and Other Evils*, 125.
9. The First International Congress of Eugenics met in London at the University of London from July 24–30, 1912. It was organized by the Eugenics Education Society of Great Britain (later the English Eugenics Society). The distinctions of those involved are obvious. The director was Leonard Darwin, a grandson of Charles Darwin. Vice presidents included: Winston Churchill, First Lord of the Admiralty; Charles Davenport, director of the Eugenics Record Office and secretary of the American Breeders' Association; Dr. Charles W. Eliot, president-emeritus of Harvard University; Dr. David Starr Jordan, president of Stanford University; and Gifford Pinchot, a pioneering environmentalist and later the governor of Pennsylvania.

> The subject of race improvement, or Eugenics, with which I have much occupied myself **during the last few years,** is a pursuit of no recent interest. I published my views as long ago as 1865, in two articles written in *Macmillan's Magazine,* while preparing materials for my book, *Hereditary Genius*.... **Popular feeling was not then ripe** to accept even the elementary truths of hereditary talent and character, upon which the possibility of Race Improvement depends. Still less was it prepared to consider dispassionately any proposals for practical action. **So I laid the subject wholly to one side for many years.** Now I see my way better, and an appreciative audience is at last to be had, though it be small.[10]

An appendix at the back of Galton's autobiography supports Galton's claim to have done so little for so long. Judging by the titles in that appendix and excluding those about genetics, at the time of his autobiography he had written some nineteen times on eugenics. Five were published during the first decade of his interest (1865–1875), when he was enthusiastic about his half-cousin Charles Darwin's new theory of evolution. That's one for every two years at a time when "popular feeling was not then ripe." Turning to other topics, during the following twenty-six years (1876–1900) he took up eugenics five times, or roughly once every five years. Finally, between 1901 and the publication of his autobiography in 1908, what he called "during the last few years," he wrote nine times, which is more than once a year (or eleven times if you count separately three articles he listed together). While his early articles focused on eugenics success (*Hereditary Genius,* 1869 and *English Men of Science, Their Nature and Nurture,* 1874), his later writings included darker and more negative themes ("Are We Degenerating," 1903 and "Restrictions in Marriage," 1905). You see that increasing pessimism in how he closed his 1908 chapter on "Race Improvement."

> This is precisely the aim of Eugenics. Its first object is to check the birth-rate of the Unfit, instead of allowing them to come into being, though doomed in large numbers to perish prematurely. The second object is the improvement of the race by furthering the productivity of the Fit by early marriages and healthful rearing of their children. Natural Selection rests upon excessive production and wholesale destruction; Eugenics on bringing no more individuals into the world than can be properly cared for, and those only of the best stock.[11]

Notice how Galton concealed the real change he wanted, which was from Natural Selection to Scientific Selection. A woman who once grieved at the death of *one* child due to natural causes, is now to grieve because, not being

---

10. Francis Galton, *Memories of My Life* (London: Methuen & Co., 1908), 310. During the middle period, eugenics was criticized within Galton's own circle, particularly in Thomas H. Huxley's 1893 *Eugenics and Ethics.* There the leading spokesman for Darwinian evolution disagreed with those who wanted to apply evolutionary ideas to human society. Unfortunately, Huxley died in 1895, removing from the scene one of eugenics most eloquent critics.
11. Francis Galton, *Memories of My Life* (London: Methuen & Co., 1908), 323.

of the "best stock," a small clique of scientists will not allow her to have *any* children. On that, Galton and Woodhull would have agreed. Where the two differed—and differed radically—was in their willingness to face controversy and ridicule. Galton avoided controversy because such behavior did not become a gentlemen of his social standing. He waited thirty-six years (from 1865 until 1901) for eugenics to acquire a small but distinguished "appreciative audience" before he acted. Woodhull, however, loved conflict and, it is easy to suspect, even exaggerated what criticism she did get. She came out for eugenics some thirty-five *before* it was fashionable. She really was the Columbus of eugenics, setting sail not knowing if her life would end in ridicule.

In the *Pall Mall* story, "Lady Eugenist," Woodhull was quoted saying, "For my part, I feel that I have done my work; I have suffered, it is true, but I have suffered for a great cause, a cause that is replete with great possibilities for the civilization of the future." Agreeing with her, the author of the *Evening Standard* article, "Eugenics in London Thirty-five Years Ago," closes with a claim that she had, "encountered a store of abuse and criticism. More fortunate than many pioneers, she has lived to see her theories accepted." That attitude must have left opponents furious. If they did not criticize, they were conceding she was right. If they did, their criticism became "abuse" and proved she was right.

Emancipation and Eugenics Linked

How about the link Woodhull saw between emancipation and eugenics? The story is not complicated, although few tell it today. (For more detail, see *The Pivot of Civilization in Historical Perspective*.) Near the end of the nineteenth century the more affluent social groups, particularly in the northeastern United States, became alarmed at their own low birthrates in comparison at the high rates of immigration and birth from the peoples of Eastern and Southern Europe. Francis Walker, Superintendent of the Census in 1870 and 1880, described the problem in 1891.

> So broad and straight now is the channel by which this immigration is being conducted to our shores, that there is no reason why every stagnant pool of European population, representing the utterest failures of civilization, the worst defeats in the struggle for existence, the lowest degradation of human nature, should not be completely drained off into the United States.[12]

Ten years later, Dr. Edward Ross, a prominent sociologist, would publish an article that provoked a long and emotional debate. Certain superior races, he said, had such "exacting standards" for their lives, that they were unwilling to make the sacrifices a large family requires. As a result, they were being supplanted by other, less able but more prolific races. His chief worry was the

---

12. Francis A. Walker, "Immigration and Degradation." *Forum* 11 (Aug. 1891), 634–44. In *Pivot*, 26–27.

LADY EUGENIST.

INTRODUCER OF THE MOVEMENT TO ENGLAND

LECTURER WHO WENT TO GAOL

To whom are we indebted for the introduction of the Eugenic idea into this country?

Somehow the public have been led to believe that Galton is responsible for the interest we now take in Eugenics, a science which only forty years ago was regarded as unfit for conversation.

But Galton can only claim credit for having, by a large sum which he left, brought the present International Eugenics Congress together.

The person, however, to whom the academic palms must fall for having been the first one to impress upon the nation the importance of race propagation on scientific principles is Mrs. Victoria Woodhull Martin, who, thirty-six years ago, came from America to England.

"Immediately after my arrival," said Mrs. Woodhull Martin to a "Pall Mall Gazette" representative, "I delivered a course of ten lectures on the subject of the scientific propagation of the race in England at the old St. James's Hall.

"I was the first lady who ever spoke on the subject in public in this country.

"My lectures caused considerable interest in England, though I was sent to gaol in America for them.

"I consider these lectures as laying the foundation of and fomenting interest in the science of eugenics.

"I worked exceedingly hard to get a hearing and I lectured, not only in London, but in Liverpool, Manchester, and Nottingham on the very subjects which are now engaging the attention of eugenists all over the world.

"Personally, I am delighted to see the congress in operation, and the subjects which it is discussing ought to engage the attention of every one who is interested in the future generation and general well-being of mankind.

"For my part, I feel that I have done my work; I have suffered, it is true, but I have suffered for a great cause, a cause that is replete with great possibilities for the civilisation of the future.

"Let the congress do the rest. I heartily wish it the success it deserves."

EUGENICS IN LONDON THIRTY-FIVE YEARS AGO.

(From a Correspondent.)

Some thirty-five years ago Mrs. Victoria Woodhull Martin amazed London by daring to deliver a lecture in the old St. James's Hall on "The Scientific Propagation of the Human Race." She was the guest of Lord and Lady Mount Temple, who had expressed grave concern for her life if she carried through her intention.

It was the first time that a woman had addressed an audience in this country on such a topic, and, indeed, the first time that a woman had spoken from the platform of St. James's Hall, if we except the occasion when Mrs. Julia Ward Howe was hooted down with hisses and imprecations. Mrs. Martin's arresting oratory won the attention of her hearers, and Prince Malcolm Khan, who was present, remarked that the lecturer was twenty-five years ahead of her time.

The views which Mrs. Martin then dared to advance have been engaging the attention of the Eugenics Congress.

"Every individual," said Mrs. Martin, "is the product of two factors—breeding and environment. If a man be an idiot or a criminal we have only to study how much is due to breeding and how much is due to environment to know the why and wherefore of his being so. Whenever you read of a crime in a newspaper you can ask yourself, 'Is that individual a victim of bad breeding or one of the martyrs of our ignorant social mal-adjustments?'"

Mrs. Martin then proceeded to suggest the remedies in words which now have become accepted truisms:—

"Remedies consist," she said, "in altering the conditions which give rise to the two classes. Improve the marriage system, and the individuals who are the product of defective breeding disappear; improve the economic conditions, and individuals who are the product of defective environment decrease. In so far as reforms have the effect of raising the standard of material well-being they will diminish or eliminate crime, drunkenness, imbecility, pauperism, insanity, and poverty. In so far as reforms have the effect of ostracising improper marriages and encouraging worthy ones, defective individuals of the first class will be exterminated."

Mrs. Martin had previously delivered the lecture from which these extracts are taken throughout America from 1870 to 1876, and had thereby encountered a storm of abuse and criticism. More fortunate than many pioneers, she has lived to see her theories accepted.

Chinese, although he also fretted about the harmful "presence in the South of several millions of an inferior race," as well as the "masses of fecund but beaten humanity from the hovels of far Lombardy and Galicia." That led him to coin a term that would become controversial.

> For a case like this, I can find not other word so apt as "race suicide." There is no bloodshed, no violence, no assault of the race that waxes upon the race that wanes. The higher race quietly and unmurmuringly eliminates itself rather than endure the bitter competition it has fail to ward off from itself by collective action.[13]

During the first two decades of the twentieth century, Theodore Roosevelt and others would pressure the nation's more well-established citizens to have more children. But Roosevelt's fumbling attempts at positive eugenics, often accompanied by praise for large immigrant families, got nowhere. In a 1917 article entitled "An Answer to Mr. Roosevelt," Margaret Sanger explained why.

> The best thing that the modern American college does for young men or young women is to make of them highly sensitized individuals, keenly aware of their responsibility to society. They quickly perceive that they have other duties toward the State than procreation of the kind blindly practised by the immigrant from Europe. They cannot be deluded into thinking quantity superior to quality.[14]

In the end, negative eugenics, targeting the poor, racial minorities, and recent immigrants, proved far more popular than positive eugenics, so much so that today within certain groups it is so deeply ingrained it does not even need to be mentioned. The basic idea was simple. From a Darwinian perspective the relative difference in birthrates was what mattered. If you are an educated woman with a career or a busy social life, it's far easier to pressure poor women to have fewer children than to have more yourself. It's even easier to let the government and social agencies do the pressuring. For a social class accustomed to regarding the poor as docile and obedient servants, it was an easy step to take.

Charlotte Perkins Gilman and Negative Eugenics

Negative eugenics was popular among post-Woodhull feminists and well-to-do women even before Margaret Sanger arose to champion birth control. You see it in the writings of Charlotte Perkins Gilman, one of the foremost feminist intellectuals of the early twentieth century. In 1914 she gave a speech that was reported in the *New York Times*. In *Pivot* I described what happened.

> Perhaps still angry about not getting flowers for Valentine's Day, in early March of 1914, Gilman launched a nasty attack on Cupid, the mythical "fat baby with a bow and arrow" at New York City's Hotel Astoria. A reporter

---

13. Edward A. Ross, "The Causes of Racial Superiority." *Annals of the American Academy of Political and Social Science* 18 (1901), 67–89. In *Pivot*, 53.
14. Margaret Sanger, "An Answer to Mr. Roosevelt." *Birth Control Review* (Dec. 1917). In *Pivot*, 157.

for the *New York Times* noted both her feminist theme ("the uplift of woman from her present state of subjection to man") and the great wealth of her audience ("200 expensively gowned examples of present-day subjection"). The half-dozen men present, the *Times* told readers, kept quiet lest "a single spark" trigger a revolution.

Cupid's fault, it seemed, was that he was male and selected marriage partners by male rules. "Man, dominant, selected woman for her sex attraction; and woman, deep in the mire of subjection, accepted." Gilman had an alternative called "Mother Love." When woman recognized her "real duty... then we can free the world of its ills." Though not completely clear in the *Times* story, Mother Love had to do with breeding a race free of social problems. Women, she assumed, wanted that, while all men wanted was a pretty face.

Opened up for questions, the subject turned to eugenics. Revealing, the only point on which all those present agreed, Gilman told her audience of well-to-do women, was "what she termed 'negative eugenics,' or the movement to prevent the entering into the world of unfit persons. The protest even against this [elsewhere], she said, was so great that one begun to believe that there were many people who could not supply the medical certificates which proposed legislation demanded."

The situation is clear. In early 1914, the stage was set for Sanger's soon-to-be movement to take birth control to the poor. Almost without exception, wealthy women (at least in New York City) were eager to see "unfit persons" (meaning poor immigrant mothers and their native-born counterparts) pressured to have few or no children. Even the ugly rhetoric that would become a trademark of Sanger's birth control movement was already in place. The opponents of negative eugenics were so stupid, Gilman claimed, that the state could bar them from marrying or having children. That's how the current hostility between the descendants of early twentieth-century Catholic immigrants and Sanger's Planned Parenthood began. It's as legitimate as hostility between black people and the Ku Klux Klan.[15]

The next year (1915) in a magazine she edited, Gilman described what she meant by "Mother Love" in a serialized novel about a feminist utopia called Herland. (You can find the eugenic-related portions in *Pivot*, or read the entire novel in a modern edition.) Here is her description of that idealized country without men where women have a way to make babies on their own.

> I understand that you make Motherhood the highest social service—a sacrament, really; that it is only undertaken once, by the majority of the population; that those held unfit are not allowed even that; and that to be

---

15. Remarks are from *Pivot*, 124. Quotations from "Cupid is Scorned by Mrs. C. P. Gilman" *New York Times* (Mar. 5, 1914), 8.

encouraged to bear more than one child is the very highest reward and honor in the power of the State.[16]

That was Gilman in 1915. When you realize that Woodhull was thinking, writing and speaking on these topics—eugenics and making a religion of motherhood—in the 1870s, over thirty years before other feminists would take up the cause, you see her importance as a pioneer. Margaret Sanger recognized that in her 1938 autobiography when she wrote, "**Eugenics, which had started long before my time,** had once been defined as including free love and prevention of contraception."[17] Woodhull had championed free love; Galton had not.

The Broad Historical Background

It might help if I offer some background to questions that are likely to arise in reading Woodhull and those who came after her.

First, there's a historical fact that many find hard to grasp. It's easy to understand why the wealthy might resent the taxes they pay for poor houses and institutions for the insane, and why they might want eugenics to purge society of such people. It's much harder, if you follow conventional wisdom, to understand why Woodhull, who was the first to publish the *Communist Manifesto* in the United States, or Margaret Sanger, who always voted for the Socialist party candidate for President (except in 1928 and 1960 when her fierce anti-Catholicism led her to vote against a Catholic candidate), would be zealous supporters of negative eugenics. After all, aren't liberals and socialists the great champions of the poor and aren't feminists the great defenders of the rights of women?

The answer hinges on how you view the world as a whole. To understand the distinction, we must go back to the French Revolution, when our current notions of a political right and left first developed.[18]

The Political Right and Left

The right is best illustrated by Edmund Burke (1729–1797) and his 1790 *Reflections on the Revolution in France*. In an introduction to the Penguin Books edition of that book, Conor Cruise O'Brien quotes from a Burke letter of March 29, 1790 in which Burke described his disdain for those who, "in cold blood can subject the present time and those whom we daily see and converse with to immediate calamities in favour of the future and uncertain benefit of persons who only exist in ideas."[19]

Let's put that in concrete terms. Imagine for a moment that you have a time machine that could travel back to 1790 and transport Burke forward to 1890.

---

16. Charlotte Perkins Gilman, "Herland." *Forerunner* 6 (June 1915), 153–54. In *Pivot*, 126.
17. Margaret Sanger, *Margaret Sanger: An Autobiography* (New York: W. W. Norton, 1938), 374. In *Pivot*, 28. Bolding added.
18. For more about what this political difference means historically, look at the writings of Jacob Lieb Talmon starting with his 1952 *The Origins of Totalitarian Democracy*.
19. Edmund Burke, *Reflections on the Revolution in France* (Penguin: Middlesex: 1969), 23.

Imagine you persuade Woodhull and Burke to go with you to visit a poor London family. You want each to tell you what should be done about the family's fourteen-year-old daughter, an passionate young women who seems a little below average in intelligence, precisely the sort of woman that Justice Holmes, in his 1927 opinion, wanted sterilized.

We know Woodhull's opinion. She campaigned across American in the 1870s, calling for something be done to keep young women like this from having their 'unfit' children, the source, she claimed, of so much crime and poverty. Woodhull would want the unfortunate young woman institutionalized until she was no longer able to have children or sterilized.

How would Edmund Burke respond? In all likelihood, he would be horror struck that Woodhull could advance such ideas "in cold blood." This woman, he would point out, wants the love of a husband and the joys of raising her own children, however impoverished her home might be. Why should that young woman, someone we can "daily see and converse with," be sterilized or forced to spend her life in the lonely wards of a female-only institution for "the future and uncertain benefit of persons who only exist in ideas." Burke would have no interest in the distant and abstract "race of gods" that Woodhull wanted to breed with present-day cruelties.

The Challenge of Thomas Malthus

For Woodhull and many like her, abstract ideals are more important than living people. They want to rid the world of abstractions such as poverty in ways that show indifference or even hostility toward the actual poor around them. You can see that in how the political left responded to Burke's contemporary Thomas Malthus (1766–1834). Today, most of those who have heard of Malthus connect his 1798 *An Essay on the Principle of Population* with the "Population Bomb" hysteria of the 1960s. But Malthus wrote for a different purpose. Unlike Burke, he fought the champions of the French Revolution with their own logic, answering their future-oriented abstractions with equally cold abstractions of his own.

At one level, Malthus' argument was brilliant. Go ahead, he wrote, create an ideal world in which no one goes hungry. It will do no good, he said. Deprived of the crude but necessary population-limiting effects of poverty, malnutrition, and disease, the number of people will eventually grow to the point where there will not be enough farm land to feed the much larger population. Population, he famously stated, grows geometrically (1, 2, 4, 8, 16...), while the food supply only grows arithmetically (1, 2, 3, 4, 5...). His argument was just the sort to impress those who prefer the abstract to the concrete, numbers to flesh-and-blood, and future ideals to present happiness.

In practice, what he said was nonsense, and that's why his predictions have fallen flat time and again. Human population can in some cases grow geometrically with a doubling time of perhaps forty years (the time it takes two adults to raise four children to adulthood). But many of our sources of food can multiply a hundred-fold in a single year, easily outpacing human population growth. While the amount of grain grown may be limited by the farm land available, no mathematical formula describes that limit. It's determined by technological advances. When Malthus traveled country roads, he saw new soil being broken by a man behind a plow horse, a slow process and one that had not changed for centuries. But today one person using a powerful machine can prepare enough soil in a single year to feed hundreds of new people. True, there is a limit to how many people the entire earth can feed no matter how advanced our technology, but it's so large and uncertain, that it's useless to inject it into present-day politics.

Charles Darwin's Answer to Malthus

Malthus did not lead the left to abandon its obsession with abstractions, but he raised questions it felt it had to answer. While it pondered, Charles Darwin's *The Origin of Species* (1859) burst upon the world. Darwin took Malthus' pessimistic argument that famines and deaths were inevitable and turned it on its head, making death an instrument of progress. The key, Darwin wrote, perhaps with an eye on the recent famine in Ireland, lay in looking at *who* was dying. You can read his argument in the very last words of *Origin*.

> Thus, from the war of nature, from famine and death, the most exalted object which we are capable of conceiving, namely, the production of the higher animals, directly follows. There is a grandeur in this view of life, with its several powers, having been originally breathed by the Creator into a few forms or into one; and that, whilst this planet has gone cycling on according to the fixed law of gravity, from so simple a beginning endless forms most beautiful and most wonderful have been, and are being evolved.[20]

Progress, grandeur, beauty, and all things "most wonderful" are, according to this famous scientist, the inevitable products of famine and death. Darwin did not solve the problem Malthus had raised. But he did allow the progress-obsessed left to modify their schemes to include a place for deaths—or methods to prevent births, which in Darwinian thought amounts to the same thing—that it had previously been so eager to eliminate. That's why Gilman's feminist utopia had to pass through a difficult stage.

> There followed a period of "negative eugenics" which must have been an appalling sacrifice. We are commonly willing to "lay down our lives" for our country, but they had to forego motherhood for their country—and it was precisely the hardest thing for them to do.[21]

---

20. Charles Darwin, *The Origin of Species* (New York: New American, 1958), 450.
21. Charlotte Perkins Gilman, "Herland." *Forerunner* 6 (June 1915), 153–54. In *Pivot*, 126.

A Biological 'Great Leap Forward'

Keep in mind that 'progress though killing' was not a new idea for the more radical left. Darwin merely added a biological rationale to an existing political one. Historically, the champions of an idealized future have proved all too willing to kill in the present to create a new world, as Burke warned and as revolutions from the Great Terror of the French Revolution to Mao's "Great Leap Forward" and the Pol Pot genocide in Cambodia have demonstrated. But most of their attention had been focused on unrepentant 'reactionaries' at the top of the social pyramid, from the nobility of the French Revolution to the educated classes in Cambodia. Darwin's new theory added a new rationale for killing, the idea that the future could be imperiled by those at the bottom of the pyramid, people who would burden the future with their inadequacies, people whose only crime lay in giving birth to children much like themselves. (That issue also underlies our current debate over legalized abortion.)

At first, it was easy to give a optimistic spin to Darwin's new idea. Charles Darwin (1809–1882) lived during the long sunset years of the British upper class. Born into a family that was already famous and acquiring a great fortune through his wife, he could live the life of a country gentleman and treat science as a hobby rather than a fiercely competitive occupation. Feminism was not yet telling the women of his class to pursue careers, and the ready availability of servants at low wages made raising a large family easy. Wealth and country life also shielded his children from the diseases of poor children in industrial cities that the poet William Blake (1757–1827) had blasted as "dark Satanic mills."

In the latter half of the nineteenth century, however, that idyllic world began to break down. Success became more competitive, forcing professional men off country estates and into crowded cities. Wealth was taxed more, making it harder to pass great fortunes on to the next generation. One man, one vote democracy became more common, weakening the power of elites. Worst of all from a eugenic perspective, women in Darwin's social class began to demand education and careers. With financial security less certain and servants more costly, the more affluent classes in England and American were having fewer children, often so few that they were not even replacing themselves, as I describe in chapters XV–XVII of *Pivot*. Equally important, improvements in public health such as clean water and food meant that famine and death no longer decimated the poor. Less than a quarter of a century after it was conceived, Darwin's great vision of inevitable human progress was in trouble.

Two Responses: Social Darwinism and Eugenics

Two responses followed. The Social Darwinians wanted to 'turn back the clock,' curtailing public health measures and returning the nation to an allegedly heroic past when only the strong survived. In chapter 9 of *What's Wrong with the World*, G. K. Chesterton referred to them as people who believed, "that the habit of sleeping fourteen in a room is what has made our England great; and that the smell of open drains is absolutely essential to the rearing of a Viking breed." Fortunately, their scheme had little chance of success. Disease is no respecter of persons. Even the wealthy benefit when the poor are no longer swept by epidemics that spread to everyone, rich or poor.

The other response was eugenics, which focused on replacing high death rates in disliked groups with low birthrates. It was just the sort of scheme to excite idealistic, future-oriented worshipers of abstractions. A superior few, experts in what Woodhull would term a "Humanitarian government" (Chapter 6), would control the wombs of the many and create a new humanity virtually free of social ills. You might call it biological socialism.

It fit well with another aspect of Darwinism. Charles Darwin had envisioned a vast biological chain of being linking man with animals that saw all as products of a complex but impersonal process. Although his theory placed man at the top, without a real rationale for doing so, it left no distinction between man and his animals ancestors. That's why so many, including Woodhull, came to the conclusion that humanity should be improved using some of the same techniques that were successful with domesticated animals. To fight that idea was, in their minds, to fight against science and progress.

This book is not intended to explore in detail those historical links or to create a new biography of Woodhull that stresses her role in pioneering eugenics. I leave that to others. Instead, I hope to stimulate a discussion of the important role that Woodhull played in the history of eugenics and to make clear what she believed by reproducing in facsimile the most important of her eugenic writings. Only few copies of the originals are available in the major research libraries of the world. This book will make them available to anyone in an easy-to-use form. A book is far more pleasant to use than a microfilm reader.

If my experience with *Pivot* is any indication, however, some readers are likely to fuss about the introductions I have written for each of Woodhull's writings. Keep in mind that those remarks are there to stimulate thought. Some will want to idolize her as a sort of secular saint, an attitude Woodhull herself encouraged. That's wrong, because much of what she said is wrong, factually, scientifically or, most important of all, ethically. Other readers will be confused by having two conflicting points of view under the same cover, mine and Woodhull's. They long for the surety of a grade-school classroom, where the right thing to do is to repeat what a textbook says back to the teacher. But life isn't a grade school; it's filled with conflicting ideas that need to be sorted out. Any time you read—and particularly when you read an advocate as single-minded as Woodhull—you should think for yourself. Woodhull has a point of view and doesn't want us to see its weaknesses. You must spot her errors yourself, and hopefully my introductory remarks will start you down that path.

Finally, some prefer to simply dismiss Woodhull out of hand, without listening to what she has to say. Like the eugenist historian I quoted, she had an undistinguished and even unsavory background and can thus be ignored. Her problems are certainly obvious, even by her own standards. She repeatedly denounced women who use sex or who marry for money, but her entire life revolved around using sex and marriage to improve her position. On the other hand, we should never forget that her early life was difficult. Pushed by her parents, who ran a fortune telling show, into marriage to an alcoholic at fifteen (1853), she had a severely retarded son named Bryon, born in 1854. Eugenics and its connection to marriage and motherhood were part of her daily life.

Using this Book

Here are a few remarks to help you make the best use of this book.

In source books such as this, some readers have trouble distinguishing the parts written by the editor from parts written by the author being republished. I have done my best to make that difference clear. All Woodhull's booklets were right-justified, so the portions I wrote have what printers call "ragged justification." You can see that on this page. The text at the end of each line doesn't end at the same place. That indicates that I wrote it rather than Woodhull. I even kept that true in Chapter 2, where I had to typeset Woodhull's text rather than use a darkened facsimile of the original. What I wrote is ragged. What she wrote is justified.

Then there is the page numbering. One of the best reasons for publishing old text in facsimile is that the result not only has the original precisely as it was published, but it lets those who use the new book reference page numbers in the original. Unfortunately, that means that books like this end up with two sets of page numbers on most pages. Fortunately, Woodhull's printers always put their page numbers at the top and centered, allowing me to place mine at the bottom outside. Keep in mind that when I make references to Woodhull's text, I use her page numbering (top center). When I make references to what I have done, I use this book's page numbering (bottom outside).

In addition, I've done my best to make the reproduced documents as easy to read as possible, even eliminating by hand most of the tiny speckles that mysterious appear when an old document is scanned. But in some cases, the original was too deteriorated to restore to a like-new appearance. For that I apologize.

Keep in mind that, although this book has grown to be almost fifty per cent larger than originally planned, it's not a complete collection of all Woodhull had to say about eugenics. It brings under one cover the articles that Woodhull herself must have considered important, as indicated by the fact that she went to great effort and expense to have five of them privately published. But additional articles on eugenics by her and others can be found in the two periodicals she published: *Woodhull & Claflin's Weekly* (New York, 1870–1876) and

The Humanitarian (London, 1892–1901). Just keep in mind something Woodhull herself pointed out. Make a clear distinction between what she authored herself. and what she merely published as an editor. Editors do not necessarily agree with all they publish.

Last of all, in this book I only lightly touch on what may be the most valuable source of information about Woodhull's role as an early promoter of eugenics. That is the impact she had on public attitudes through the many speeches she made across the United States, as described in local newspaper articles written after she visited a town. The influence she had on affluent, influential women through those speeches and newspapers may be of particular importance in explaining the later spread of eugenic ideas among women in the United States after 1900. It was Woodhull, after all, who first promoted, widely and publicly, ideas about improving society by making better children through science and planning that Charlotte Perkins Gilman and Margaret Sanger would later exploit. As I said at the start of this chapter, too many historians have followed the lead of later eugenists and overstressed the role of prominent, credentialed men such as Francis Galton, while neglecting the role women such as Woodhull had in the early history of eugenics.

I've included extracts from some of those newspaper articles, particularly in Chapter 3, but most of those I include were selected by Woodhull herself. From them, it is possible to see that Woodhull was the target of criticism, but it is difficult to get an accurate understanding of just what her critics were saying from reporters who so obviously adore her. That problem can easily be remedied. This book and the roughly 300 pages of "Press Notices" in her 1890 *The Human Body the Temple of God* give the date and place of dozens of her speeches, so it should not be hard to track down other articles that take a more objective view of what she was saying and how her remarks were regarded. There are also articles on eugenics (although it did not yet have that name) in *Woodhull and Claflin's Weekly* (New York, 1870–1876), as well as in *The Humanitarian* (London, 1892–1901). If this book had not already grown large, I would have included more of them here.

I would be delighted by such research. This book is not intended to have the 'last word' on Victoria Woodhull's role in the early spread of eugenics. Instead it hopes to have a 'first word' on her role as the pioneering Lady Eugenist, in the hope that would encourage others perhaps more gifted and industrious than I to explore the topic in greater depth.

Chapter 2
Children—Their Rights and Privileges
Introduced by Michael W. Perry

The 1871 speech that is republished in this chapter demonstrates that Victoria Woodhull was promoting eugenics in the 1870s just as she claimed. That is obvious as early as the second paragraph.

> I propose to speak briefly of children—a subject which, though comparatively ignored, is to me one of the most important. I believe that Spiritualists have an interest in all kinds of reform; and therefore, must have in this, which lies at the basis of all others, since a perfected humanity must come of perfect children.

Notice Woodhull's claim that, "a perfected humanity must come of perfect children." That means serious intrusions into family life, a fact she knew when she demanded that society use "all the means" used to perfect "things"—animals, plants and perhaps machines. That's the essence of eugenics.

> It is laid down as an undeniable proposition, that the human race can never even approximate to perfection until all the means of which men make use to produce perfect things are also made use of in their own production.

Unfortunately, Woodhull's published version cannot be reproduced. The article in *Woodhull & Claflin's Weekly* for October 7, 1871 (3–5), available on microfilm, is too dark. Instead, a carefully proofed text is included below.

Woodhull's article stated that the text came from the "Original Report of the Eighth National Convention—The American Association of Spiritualists" edited by Henry T. Child. Woodhull delivered the address on the afternoon of September 12, 1871, the first day of the convention. It was originally entitled, "The Training of Children—Good Advice to Mothers," but has since become "Children—Their Rights and Privileges," from a catch phrase she used in the speech. That shifted the focus from mother to child and was more in keeping with the spirit of an article that's hostile toward mothers training their children. Woodhull, for instance, made this remark near the close of her speech.

> To make the best citizens of children, then, is the object of education, and in whatever way this can be best attained, that is the one which should be pursued, even if it be to the complete abrogation of the present supposed rights of parents to control them.

As you read, notice what Woodhull is saying. Parents rights are "supposed," while those of the State are absolute. For her, both parent and child "belong to humanity"—meaning the government. Her goal was *not* to free a few abused or neglected children from exceptionally bad parents. It was to have the government replace parents "fully one half" the time or more.

> ... For ourselves we make the distinct assertion that we are thoroughly convinced that fully one half the whole number of children now living between the ages of ten and fifteen, would have been in a superior condition—physically, mentally, and morally—to what they are, had they been early entrusted to the care of the proper kind of industrial institutions.

Remarks like those illustrate why, like Margaret Sanger, Woodhull moved easily between radical socialist circles and friendship with her era's most powerful industrialists. Both regarded humanity as raw material to be shaped by their betters. Eugenics would find its greatest champions among two groups not often linked—socialists and the wealthy. Thanks in part to Woodhull, by the early twentieth century, some feminists were also playing a major role in eugenics. Regimenting other women, particularly women thought to have too many children, seems to have had a particular attraction for women with few or no children. That is illustrated by the forced labor camps that Charlotte Perkins Gilman advocated in her 1908 "A Suggestion on the Negro Problem."

> But the whole body of negroes who do not progress, who are not self-supporting, who are degenerating into an increasing percentage of social burdens or actual criminals, should be taken hold of by the state....
>
> ... Construction trains, carrying bands of the new workman, officers and men, with their families, with work for the women and teaching for the children, would carry the laborer along roads he had made, and improve the country at tremendous speed.[1]

Gilman claimed her scheme was "not enslavement but enlistment," but fathers laboring at jobs they did not choose, and the mothers separated from their children by forced labor would have felt otherwise. For both Woodhull and Gilman, the State decides what constitutes a useful citizen and can do what it wants to create one. Near the end of her 1871 speech Woodhull described with dreadful finality the absolute power of the State. Parents who object, she believed, are unaware just how poor they are as parents or are too lost in "selfishness and "mere promptings of affection" to see "the good of the child." For Woodhull the best parents were passive, not standing between the State and their child and not interfering with the future it intends to create.

> To make the best citizens of children, then, is the object of education, and in whatever way this can be best attained, that is the one which should be pursued, even if it be to the complete abrogation of the present supposed rights of parents to control them. It is better that parents should be able to look with pride upon their children grown into maturity, as youthful citizens by the assistance of the State, having been unable to make them thus themselves, than to consult the sentiments of the heart, by having them con-

---

1. Michael W. Perry, ed. *The Pivot of Civilization in Historical Perspective* (Seattle: Inkling Books, 2001), 123–24. The book documents the hostility women from social groups that had few children often felt for women from religious or immigrant groups with large families.

stantly under their care; and by so doing allow them to grow into maturity in form and grace, yet lacking the necessary elements to make them acceptable to, or to be desired by, society. One of these is the result of the existence of wisdom; of affection, guided by reason; the other that of selfishness in which the good of the child is sunk in the mere promptings of affection, regardless of consequences. No reasonable person can question which of the two is the better for all concerned for children, for parents and for society.

In the decades that followed, many would find those claims compelling. And, although what Woodhull meant by the words that follow is uncertain, there was menace in the close link she saw, near the end of her speech, between "the rights and privileges of children" and having "their true relations to society... properly enforced."

> We trust that the time is near when the rights and privileges of children will be duly accorded and guaranteed to them by society, and when their true relations to society will be scientifically analyzed and understood and properly enforced.

Perhaps the best way to understand that remark is that children will have the "rights and privileges" to be the sort of people the State wants them to be and no right to be anything else. They will be "perfect" or else.

Also, remember that during the latter half of the nineteenth century spiritualism was much more fashionable than today. The fact that Woodhull gave this speech at a spiritualist convention did not marginalize what she said. Spiritualism was viewed by many as a bold new faith that was sweeping away the cobwebs of traditional religions. Only later would spiritualism acquire a reputation for deception and fraud. The fact that Woodhull turned to spiritualists is also an early indication that the chief foes of eugenics would be conservative Catholics and Protestants, those she described here as defending, "hoary-headed bigotry, blind intolerance and fossilized authority."

Most important of all, notice that there are no references to early English champions of eugenics such as Francis Galton. Woodhull's arguments come almost entirely from everyday life. In a nation that was still agricultural, her desire to breed people like horses or tomatoes would have been readily understood, however much some might object for religious or other reasons. Galton's arguments, in contrast, were based on how often fathers and sons were listed in books of eminent citizens.[2] Those from privileged English classes understood Galton. Ordinary Americans would find Woodhull's farm-based arguments easier to follow. (Bolding added to Woodhull's text.)

The complete text of Victoria Woodhull's 1871 speech to the spiritualist convention follows.

---

2. Portions of Francis Galton's original argument are reprinted in G. K. Chesterton, *Eugenics and Other Evils* (Seattle: Inkling Books, 2000), 123–126. The original was published in *Macmillan's Magazine* in 1865.

Children—Their Rights and Privileges
by Victoria Woodhull

I scarcely know how it has come about that I am on this rostrum, in the midst of a Spiritualistic convention. I have been a Spiritualist and a recipient of heavenly favors ever since I can remember, but for reasons never explained, I have not been known to Spiritualists, nor they to me. In my humble way, I have been an earnest advocate of the principles of the spiritual philosophy, while to me its truths are quite as real as are the facts of material existence; and all my hopes for the future of humanity are founded upon the inauguration of a complete unity of purpose between the two spheres in all things upon which the good of humanity depends. I thank this convention for its hand of fellowship when so many others are set against me. If I have faults and errors, they have come from a misunderstanding of Him to whom I owe all that I am and who in my childhood taught me of the angels, in my youth smoothed the stony paths I trod, and in maturer years instilled in my heart a love for all humanity, and to be whose servant is still all my ambition.

I propose to speak briefly of children—a subject which, though comparatively ignored, is to me one of the most important. I believe that Spiritualists have an interest in all kinds of reform; and therefore, must have in this, which lies at the basis of all others, **since a perfected humanity must come of perfect children.**

We have often wondered that, among all the medical authorities, there have not been more who devoted some part of their profuse writings to the ante-natal care and treatment of children. No more important addition could be made to our system of social economy, nor to our pathological literature, than a strict analysis of foetal life for popular circulation. While so much has been said and written regarding children's care and treatment after birth, that part of their life previously has been entirely ignored. It would be just as proper to ignore their life after birth until some still future period, say three, five or seven years of age, as to do so before birth.

To lay a good foundation for a good life, it is required that the proper care should be bestowed upon it from its very point of beginning. The tiller of the soil exercises special care and his best wisdom in preparation for the future harvest. **He knows, from oft repeated experience, how important it is to have the very best seed, of the very best variety.** He knows that seed thus selected, planted side by side with unselected seed, and receiving no more care, will yield not only larger harvests but also that they will be of choice quality.

Having the best seed possible, his next step is to have the ground properly prepared, into which at just the proper season, he deposits it. All these preparatory measures are a part of the process by which our fruits, grains, and vegetables have been brought to their present state of perfection. Everybody knows that fruits and vegetables which grow wild are poisonous, [but] are capable of being brought by cultivation to be delicious articles of diet. **Everybody knows** that by study and care our most celebrated breeds of horses and other stocks of domesticated animals have been obtained. Everybody knows that deep scientific research is constantly being made in almost every department of production, and that those engaged in the respective departments eagerly apply every new fact which science makes clear. It is an admitted fact that the future character of what is to be produced, can be very nearly, if not absolutely determined by those who have charge of the process. Even the color which the herdsman desires for his cattle can be obtained; and what is true regarding color is just as true regarding all other indices of individuality.

Notwithstanding all these accepted facts which are coming to be the rules and guides of people, **when we approach the subject of making the same rules and guides so general in their application as to include children—the world stands aghast and with one united effort, frowns it down.**

Nobody denies the importance of the subject, but those who speak at all argue that it is one of those things which we are not prepared to meet. Not prepared to meet! And the whole Christian world has been preaching regeneration these eighteen hundred years! which they tell us is the one thing necessary. All the importance claimed for regeneration we willingly admit; all badly produced persons require regeneration; but as to it being the main thing, we beg to demur. If regeneration is an important matter, generation is still more so. It is to the consideration of this fact, as demonstrated and practiced in all departments of nature below, that the human must come and acknowledge itself a proper subject of. Just so far as science can demonstrate and humanity will put its demonstrations to practice, just so far can the necessity for regeneration be done away.

It is too true that the courage to face this question is generally wanting and when it is attempted, all society pretends to be outraged. **Are human beings, then, to be always considered of so much less importance than the things they make subservient, that they should forever be left to come into this world's existence as individuals at random?** We know the obloquy that has fallen upon all who have ever attempted to hold the mirror so that society would be obliged to contemplate itself; but, notwithstanding all this, we feel there is not a more noble object. We have considered all the bearings of this matter, and have determined to stand by the flag we have thrown to the world:—"Children their Rights, Privileges, and Relations," and we shall maintain it argumentatively, if possible; defiantly, if need be, against all opposition, let it come from whence it may, or let its character be what it may. Argument we know we shall not have to encounter. Scientific hindrances we know we shall not find in our path. Common sense we know will offer no word of reproof. **We shall, however, encounter hoary-headed bigotry, blind intolerance and fossilized authority—and we are prepared.**

It is laid down as an undeniable proposition, that the human race can never even approximate to perfection until all the means of which men make use to produce perfect things are also made use of in their own production. Let those who decry the proposition turn to their so much revered Bible and read: "Ye cannot gather figs of thorns nor grapes from thistles"—and learn wisdom therefrom. It must be remembered how great an "Infidel" was he who first demonstrated Arterial and Venous Circulation, which has come to be of such importance in diagnosing diseases. It is generally true that those things which result in the greatest benefit to humanity, meet with the most blind and insane opposition in their first struggles for recognition. If this subject of children is to be judged by this rule, it is to develop into greater importance than any which has yet occupied the human mind.

But it is said, how can this be done? It cannot be done immediately to the fullest extent, but the recognition of its importance can be forced upon humanity, and the practice of its evident deductions can be attained by degrees. Once let it become divested of this absurd idea of "impropriety," and humanity will begin to practice its teachings. It is only required that reason be exalted to its proper place and influence, and analogies, with which nature abounds, will become the great teachers.

The difficulty with which we shall be met at every step is, that it is nearly impossible to make people realize that their lives here are for any other or higher purpose than for each of them to acquire for him or herself the greatest amount of personal gratification. **They cannot yet sufficiently realize that each individual is made one of the means by which the whole of humanity is advanced.** They cannot yet be brought to reduce to practice what all admit, that he or she is the greatest man or woman who does the most for humanity; nor have they more than an undefined belief that in doing the most for humanity, they do most for themselves. Yet this has been the logic of the doctrine of Christianity nearly two thousand years.

The teachings of Christianity are well enough, they have been taught persistently. But we have now arrived at that age of the world which demands adequate results as proofs of the validity of assumed positions. The Apostles taught that "certain signs" should follow those who believed. Do these signs exist within the heart of the professed representatives of true Christianity? By their fruits shall ye know them. We do know them by their fruits, which are not so perfect as to warrant the conclusion that humanity has passed from being "professors" into being "possessors."

Human life may be compared to a military campaign, in which no amount of valiancy and good generalship can overcome the defects of an imperfect organization of the "body" with which it is to be made. We may as consistently expect a badly organized army to make a good military campaign as to expect a badly organized child to make a good social campaign. To this the very beginning of organization should all reformers turn who expect to produce any beneficial results, which shall be ultimate and lasting, and which shall mark the perfecting process of humanity.

Women by nature, are appointed to the holy mission of motherhood, and by this mission, are directly charged with the care of the embryotic life, upon which so much of future good or ill depends. It is during this brief period that the initials of character are stamped upon the receptive, incipient mentality, which, expanding first into childhood and on to manhood or womanhood reveals the true secrets of its nature.

The rights of children, then, as individuals, begin while yet they are in foetal life. Children do not come into existence by any will or consent of their own. With their origin they have nothing to do, but in after life they become liable for action which perhaps was predetermined long prior to their assuming personal responsibility. In youth, children are virtually the dependencies of their parents, subject to their government, which may be either wise or mischievous, and is often the latter as the former. But having arrived at the proper age, they step into the world upon an equality with others previously there. At this time they are the result of the care which has been bestowed upon them from the time of conception, and whether they are delivered over to the world so as to be useful members of society, or whether they go into it to prove a constant annoyance and curse, seems to be a matter which cannot be made into such personal responsibility as to make it a subject of their own determining. At this period they find themselves possessed of a body and a partially developed mind, in the union of which a harmonious disposition and character may have resulted. Respectively, they are possessed of all shades of disposition and character, from the angelic down to the most demoniacal; but all these are held accountable to the same laws; are expected to govern themselves by the same formula of associative justice, and are compelled by the power of public opinion to subscribe to the same general customs.

All people are obliged to meet the world with the characteristics with which they have been clothed, and which they had no choice in selecting. When all things which go to make up society are analyzed and formulated, it comes out that society holds its individual members responsible for deeds of which it is itself indirectly the cause, and therefore responsible for.

It is a scientifically demonstrated fact that the mind of every individual member of society is the result of a continued series of impressions, which are continually being received by their senses, and transmitted to and taken up by consciousness, which becomes the individuality of the person. If any one doubt this, let him listen to what Prof. J. W. Draper, President of the New York Medical University College, says upon this subject:

> There are successive phases… in the early action of the mind. As soon as the senses are in working order… a process for collecting facts is commenced. These are the first of the most homely kind, but the sphere from which they

are gathered is extended by degrees. We may, therefore, consider that this collecting of facts is the earliest indication of the action of the brain, and it is an operation which, with more or less activity, continues through life.... Soon a second characteristic appears—the learning of the relationship of the facts thus acquired to one another. This stage has been sometimes spoken of as the dawn of the reasoning faculty. A third characteristic of almost contemporaneous appearance may be remarked—it is the putting to use facts that have been acquired and the relationships that have been determined.... Now this triple natural process... must be the basis of any right system of instruction.

It appears, then, that contact and constant intercourse with external manifestations is not only necessary for the production of thought and its collaterals, but that to retain the consciousness which makes thought possible such manifestations must be continuously impressed upon the individual. This seems to be conclusive that mind is the result of the experiences of the manifestations of power.

Without these experiences, children would grow up simply idiotic. The "Professor" says emphatically, that a recognition of this process must be the basis of any right system of instruction. To state the proposition comprehensively, the education of children should consist in surrounding them by such circumstances and facts as will produce upon them those effects which will tend to develop them toward our highest idea of perfect men and women.

The chief difficulty about these things is that their direction has been assumed by the professors of religion rather than by scientists. Science is eminently progressive; religion is as eminently conservative. Science, in its analysis of the facts of the age, comes in direct conflict with the theories of religious sects. Happily, these things are now undergoing change, and they who once taught that the world was created out of nothing in six days and nights, of twenty-four hours each, have given way to the demonstrations of geology, and are forced to admit that their previous belief was founded in an allegory.

The common practice of the world, in all things which it desires to modify or remedy, is to begin at the extreme, where the effects are found, and from them to work backward toward the beginning. The whole course of the world regarding crime has been to punish rather than to prevent it; to work with the effects of education. What men or women are at the time they become recognized citizens, society makes them. They are its production, as much as the apple is the production of the tree. If the apple is a bad apple, it is not its fault; that lies in the tree. **If men and women are bad men and women when they arrive at legal age, it is not their fault, but it is the fault of society in which they are born, raised and educated.**

It is scientifically true that the life which develops into the individual life never begins. That is to say, there is no time in which it can be said life begins where there was no life. The structural unit of nucleated protoplasm, which forms the center around which aggregation proceeds, contains a pulsating life before it takes up this process. The character of the nerve stimula of which this is possessed and which sustains this evidence of life, must depend upon the source from which it proceeds. **In other words, and plainly, the condition of the parents at the time of the conception is a matter of prime importance, since the life principle with which the new organism is to begin its growth should be of the highest order.**

Cases of partial and total idiocy have been traced to the beastly inebriation of the parents at and previous to conception. On the other extreme, **some of the highest intellects and the most noble and lovable characters the world ever produced, owed their condition to the peculiarly happy circumstances under which they began life,** much of the after portion of the growing process of which having been under favorable circumstances. Many mothers can trace the irritable and nervously disagreeable condition of their children to their own condition at this time.

We are aware that these subjects are almost unanimously ignored by society; also that society pretends to blush at the mention of them; and well it may blush, for the abortions of nature which it is continually turning upon the world to be its pests, its devils, its damnation and their own worst enemies are sufficiently hideous to make all humanity blush with well-founded shame.

But the time must come wherein they will not only be discussed, but when a full knowledge of what pertains to conception, foetal life, birth and growth to full manhood and womanhood will be an important part of every child's education.

Virtue nor modesty does not consist in the avoidance, the ignoring or ignorance of these things, but true virtue, true modesty and true general worth consist, in part, at least, of a complete knowledge and practice of them. It is full time that we have done with the sham modesty and affected virtue with which humanity has been cursed.

It is required that we begin at the very root of the matter, and that lies in the condition of persons about to become parents. And just to this point is where the woman question leads. It is the important question of the age, and it will rise to be thus acknowledged. **All present humanity has a direct interest in it, and all future humanity demands of the present its right to the best life which it is possible to have under the best arrangement of present circumstance which can be formulated.** And there are those who will not permit that their rights be much longer ignored. There will be "John the Baptists" preaching in the wilderness, "Prepare ye the way," and humanity must and will heed them. Such is the prophecy of the present; and the present will do well to listen to its teachings.

The *New York Tribune* asserts that the cause of half the vice among us is the ignorance of parents of the fact that certain nervous and cerebral diseases transmitted from themselves tend to make of their children from their birth criminals or drunkards and that only incessant and skillful care can avert the danger. The editor then goes on to philosophize in this way:

> A man may drink moderately but steadily all his life, with no apparent harm to himself, but his daughters become nervous wrecks, his sons epileptics, libertines, or incurable drunkards, the hereditary tendency to crime having its pathology and unvaried laws, precisely as scrofula, consumption, or any other purely physical disease. These are stale truths to medical men, but the majority of parents, even those of average intelligence, are either ignorant or wickedly regardless of them. There will be a chance of ridding our jails and alm houses of half their tenants when our people are taught to treat drunkenness as a disease of the stomach and blood as well as of the soul, to meet it with common sense and a physician, as well as with threats of eternal damnation, and to remove gin-shops and gin-sellers for the same reason that they would stagnant ponds or uncleaned sewers. Another fatal mistake is pointed out in the training of children—the system of cramming, hot-house forcing of their brains, induced partly by the unhealthy, feverish ambition and struggle that mark every phase of our society, and partly for the short time allowed for education. The simplest physical laws that regulate the use and abuse of the brain are utterly disregarded by educated parents. To gratify a mother's silly vanity during a boy's school days, many a man is made incompetent and useless. If the boy shows any sign of unnatural ambition and power, instead of regarding it as a symptom of an unhealthy condition of the blood vessels or other cerebral disease, and treating it accordingly, it is accepted as an evidence of genius, and the inflamed brain is taxed to the uttermost, until it gives way exhausted.

When a paper, which so religiously ostracizes so much which is involved in the principles of general reform, as the *Tribune* does, comes so near to the "root of the matter," it may be seriously considered whether the time has not arrived in which to speak directly to the point. **The remedy is twofold: first, and mainly, to prevent the union of persons addicted to false practices; second, to endeavor to reform those who are already united.**

A positive assertion is here made. **No two persons have the right to produce a human life and irremediably entail upon it such a load of physical and mental hell as the *Tribune* cites.** It is the merest sham of justice to punish the drunkard for the sins of his or her parents. It is the most superficial nonsense and the purest malice to curse the bad fruit which grows in your orchard because you do not take care of the trees; but it is no more so than it is to curse and punish children for the crime of their parents.

Marriage or the union of the sexes is a natural condition of the human race. Whatever relations they may sustain to the children they produce, those which society as a whole sustains to them are broader and more comprehensive. The parents are but parts of society, and their children are nothing less, so that while they, by present social systems, are for a long time left to the special control and guardianship of their parents, it can be considered only as in trust for society.

The relations which should be considered as the foundation of society are those which exist between society and marriage in its special function of reproduction, which thus far has been utterly ignored. When two are about to form a marriage union, does society in its legitimate functions of promoting and protecting the public welfare ever stop to ask what the results of the union are likely to be? Instead of this question entering into the consideration, the only one that has been thought of is: How shall these two be compelled to live out the remainder of their natural lives together, utterly regardless of the higher thought of children?

It is a well-established fact among the medical profession that nearly all the consumption which hurries so many victims through life has its source in hereditary syphilitic taint, which for delicacy, has been christened scrofula. **Now what business or right has a man or woman, who knows that his or her system is loaded with this infernal poison, to become the propagator of the species? The same is equally true of all other diseases and damnations which can be transmitted, and not more of those which pertain to the purely physical than of those which relate to the mental and the moral.** When the world shall begin to act upon this deduction, it will have commenced a course of advancement which will never be intermixed with retreats.

Education in matters which refer to these vital points should be one of the first steps to be taken by society. They have been foolishly and criminally ignored upon the false premises that to instruct children in them would be to lead them into unfortunate conditions, whereas the very reverse is the truth. If there are dangers to be avoided, the very best way to prepare children to avoid them is to give them a perfect understanding of what they are. In knowledge there is always safety. In ignorance there is always danger.

Let these truths be adopted in the education of children, regarding their duties as the future parents of society, and one-half the ills with which society is inflicted would soon disappear.

If our houses of prostitution were searched and their inmates questioned, none would be found whose mothers had the good sense to teach them the objects and functions of their sexual systems. It is the ignorance of these things which fills these blotches upon the fair face of humanity.

There is a law common to nature by which those things that are best adapted to each other are brought and held together. There is a chemistry of the social, intellectual and moral sentiments as well as of the material elements. Education should include a perfect knowledge of this chemistry, so that compatibles may be apparent at once to all people of both sexes. Open the fountains of knowledge, so that all may drink of the waters of a true life.

Children, by the little things they so readily gather about the difference of sex, are made curious to just the extent the means of satisfying that curiosity is difficult, and they pursue their means by stealth whenever and wherever possible. This results in producing a morbid condition of the mind about it, and encourages all kinds of secret vices, which are sapping the very life and beauty of the coming generation. No one can doubt this who will give it the attention it merits, to be one of the crying evils of present systems of education. If instruction were begun in these matters at or about the age when curiosity is developed, and it is made a matter of course, is it not plain that it would produce effectual results?

We are aware that "conservatives" will decry this plain way of treating this subject, and make use of the usual method of manifesting their condemnation; nevertheless, the proposition to us is a simple one, over which we have spent many weary hours. A secret attracts everybody's attention. When it is a secret no longer it ceases to attract attention, and becomes reduced to its legitimate and natural uses. We assert our belief the same results would follow the education of children in sexual matters; knowledge would succeed curiosity, and healthy action of the mind a morbid desire.

We now approach a part of the subject which is of supreme moment, and that is the care which embryotic life demands. During this period, every influence to which the mother is subjected, be it ill or good, produces its effect upon the embryo. Whoever is an adept in these matters can go through society and from each individual tell what circumstances his or her mother was surrounded by during her pregnancy. Mothers of humanity! Yours is a fearful duty and one which should in its importance lift you entirely above the customs of society, its frivolities, superficialities and deformities, and make you realize that to you is committed the divine work of perfecting humanity.

Under our systems the interests of children are utterly ignored. No matter how illy-mated people may be, children will result. It will be difficult to find a case, even where actual hate exists, and not find children. What can be expected of children generated, born and raised under such influences. There are numerous influences constantly being made public where mothers are even brutally treated during pregnancy, and oftentimes because they are pregnant.

Just the life the mother leads will she prepare her child to lead. Just what the mother desires to make her child she can mould and fashion it to be. What a condemnation these considerations are upon the practices of fashionable society. **How utterly worthless are the lives of so many mothers, and how devoid of purpose. Just so are their children.** In the insane desire for dress and display, which characterizes so many women, lies the bane of life for their children. The cold heartlessness of the woman of fashion contains the germ of destruction for her daughter and the seeds of vice for her son. No warm-hearted, generous-souled children can spring from such soil.

So also is abortion a practice which spreads damnation world-wide. Not so much, perhaps, in those cases where it is accomplished, but in those much more numerous cases where it is desired and attempted, but not reached. When a woman becomes conscious that she is pregnant, and a desire comes up in her heart to shirk the duties it involves, that moment the foetal life is the unloved, the unwished child. Is it to be wondered that there are so many undutiful children—so many who instinctively feel that they are "encumbrances" rather than the beautiful necessities of the home?

What true mother's heart but bounds with pride and joy when she sees the beauteous results of her constructive work? Why should she not also feel happiness when she realizes she is performing that constructive process? Is it to be wondered that there are so many children lacking all confidence in themselves and so foolishly diffident that it follows them through life, when we consider the conduct of women during pregnancy? It should be the pride of every woman to be the willing, the anxious, the contented mother, and if she be so under the guidance of the knowledge we deem essential, she will never have cause to regret that she fulfilled the duties of maternity. All practices which degenerate the character of children should be discountenanced by every humanitarian, and women encouraged to wisely and perfectly mould and fashion the life which they shall give to the world.

But we must pass from ante-natal life to that which has so generally been considered the beginning of it, and here a searching examination develops little more to be approved than found previously. How little scientific or acquired knowledge there is in regarding the early care of children—their immense death rate clearly shows. It seems one of the most sorrowful things of life to see the merest babes drop off by thousands for the reason that mothers do not know how to rear them.

If wives will become mothers without the knowledge requisite to fit them to perform their duties to their children then they should themselves be put under the care of some competent authority, so that the life they have been instrumental in organizing may not be uselessly thrown away. We are arguing, pleading, urging the rights of children; those rights which shall make every child, male and female, honorable and useful members of society.

Whether in acquiring this right, all old forms, all present customs, all supposed interests are found standing in the way, matters not, **the question is, "What is for the best interests of children, not merely as children, but principally as the basis of future society?"** Scarcely any of the practices of education, of family duties or of society's rights in regard to children are worthy of anything but the severest condemnation. They do not have their inherent rights at all in view. They consult the affections to the exclusion of all reason and common sense. They forget that the human is more than an affectional being; that he has other than family duties to fulfill, and that he belongs to humanity, which is utterly ignored by all present practices.

Let the father and mother of every family ask themselves: Are we fully capable of so rearing our children that no other means could make them better citizens and better men and women? And how many could conscientiously give you an affirmative answer? The fact that children are born and grown to be citizens, and not to remain children of the parents simply, is overlooked.

We are aware that this, if intended for any considerable and comprehensive application, would be regarded as a startling assertion. Many true things when first announced startle the world which thought differently so long. **For ourselves we make the distinct assertion that we are thoroughly convinced that fully one half the whole number of children now living between the ages of ten and fifteen, would have been in a superior condition—physically, mentally, and morally—to what they are, had they been early entrusted to the care of the proper kind of industrial institutions.**

We hold it to be an absolute and fundamental right that every child female and male, has, that when they are received into society as determining powers, they shall be possessed of the required capacity and experience to take care of themselves, and to perform what may be required of them. Those who are best prepared to fulfill the duties which can by any possibility devolve upon them as members of society, are the best citizens, and give unanswerable evidence of having been the recipients of the best means of growth and education.

To make the best citizens of children, then, is the object of education, and in whatever way this can be best attained, that is the one which should be pursued, even if it be to the complete abrogation of the present supposed rights of parents to control them. **It is better that parents should be able to look with pride upon their children grown into maturity, as youthful citizens by the assistance of the State, having been unable to make them thus themselves, than to consult the sentiments of the heart, by having them constantly under their care; and by so doing allow them to grow into maturity in form and grace, yet lacking the necessary elements to make them acceptable to, or to be desired by, society.** One of these is the result of the existence of wisdom; of affection, guided by reason; the other that of selfishness in which the good of the child is sunk in the mere promptings of affection, regardless of consequences. No reasonable person can question which of the two is the better for all concerned for children, for parents and for society.

The weight of our proposition, that society is itself responsible to children for the condition in which they are admitted to it as constituent members of itself, must begin to be apparent, for so far as they are concerned up to that time they are not responsible. This being self-evident, is it not also self-evident that they can not with any consideration of justice, be held to account for that which is the legitimate consequences of, and which is positively determined by, that condition?

We trust that the time is near when the rights and privileges of children will be duly accorded and guaranteed to them by society, and when their true relations to society will be scientifically analyzed and understood and properly enforced.

Then will the prophecies of all ages have reached the consummation; then will commence the earthly reign of the King of kings and Lord of lords, as prophesied by all the holy prophets of the world; then old things shall pass away, and all things become new; then The Christ shall sit upon the throne, and from his inexhausted fountain of love, justice shall continually flow over all the earth, "as the waters cover the sea."

As vanish the heavy mists of the morning before the radiance of the rising sun, so will vanish the clouds that hang around the minds of men, and shut out the rising spiritual sun, for whose "star in the East" wise men are constantly watching; the sun that will rise higher and higher, and extend its rays wider and wider, until it shall enlighten the minds of all mankind, until the icebergs of ignorance, tradition, and superstition are dissolved which now float in the ocean of progress—society with its cankered, festering heart; commerce robbed of its legitimate function; labor of its recompense, and religion of its spirituality, education lacking wisdom, marriages forming disunions, and women without rights.

All the false forms of the present must yield their sway to God's command—"Let there be light." The laws of God are never changed—though old as creation, they are ever new, ever sufficient for all the vicissitudes of life; they are ever full of wisdom, justice, and love; they are written all over the face of creation, in the bosom of the earth, and in the heart of man; they are uttered by the raging tempest that rocks the mighty ocean; in the terrible mutterings of the earthquake, in the fury of destructive battle, when hosts are hurled on hosts in fratricidal strife; through all these the voice of God proclaims—"Let there be light," and there is light.

We also hear its whispers in the gentle zephyrs that stir the bursting buds, and in the blooming flowers that lift their heads to drink the falling dew; in the hum of busy nature; in the gushing fountain; we see it in the gambols of the bubbling brook; in the mother's love for the new-born life; in the father's pride; in the unspoken joy of the maiden's soul, listening to the first sweet tones of love; in the magneticities of human sympathy which bind all mankind in a common brotherhood and in the dawning light of heaven brought to earth by the angelic hosts to usher in the reign of universal, justice, peace and love."

At the conclusion of Mrs. Woodhull's remarks, Dr. H. B. Starr of Boston, offered the following resolution: "Resolved, That this Convention is honored by the participation in its deliberations of Mrs. Victoria C. Woodhull, whose wise selections of the fundamental subject of reform has been fully justified by her able statement of its importance, and that our thanks are hereby expressed to her for her comprehensive plainness of speech and true delicacy with which this eminently radical subject has been treated by her."

The resolution was unanimously adopted.

Chapter 3

Press Notices

Introduced by Michael W. Perry

> *The great demand of the age is for better men and women. But here comes a woman, ready to tell you out of the fulness of a mother's heart how to bring into this world better men and women, and you start back with horror!*

When she published *The Human Body The Temple of God* in 1890, Victoria Woodhull intended to put in permanent form what she had said in hundreds of speeches made across the United States and the United Kingdom between 1869 and 1882. In the Preface, she noted the twin purposes of her American speeches. "The topics I dealt with were mainly two: the one political, the other social. The question of Female Suffrage, and of woman's political rights under the Constitution of the United States, was then, and still is, of interest to many; the question of raising society, through woman, to a higher standard of morals should be of supreme moment to all."[1]

Notice the contrast between "Female Suffrage" being "of interest to many" and "raising society"—her broad eugenic scheme—being "of supreme moment to all." Make no mistake, Woodhull believed that her ideas about eugenics were more important than winning the vote for women. In a 1876 speech in New Jersey she went even further, a reporter writing that she, "declared that it was useless to discuss suffrage until the women of the country had raised up a better race of man."[2]

That quote is but one example of newspaper reports about Woodhull's speeches that fill over 300 pages of *Human Body* and form a collection of "Press Notices of Extemporaneous Lectures Delivered Throughout America and England from 1869 to 1882." They make clear that her influence was wide-ranging and deep. The cities in which she spoke ranged from Davenport, Iowa and St. Paul, Minnesota to Rutland, Vermont and included major cities such as Chicago, Detroit, and even snobbish Boston, where she was long unwelcome.

In a speech in Omaha during early 1874 she went even further in stressing the importance of her social agenda, claiming that "negro slavery was not so great a cause for dissatisfaction then as are the more subtle slaveries now."[3] In speech two weeks later, those more subtle modern slaveries were linked to her demand that daughters be allowed to "marry for love and not for money or homes."[4] Always keep in mind that in Woodhull's early form of eugenics "love" had a magical power to produce superior children. It was the bridge between

---

1. Victoria Woodhull, *The Human Body* (London: Hyde Park Gate, 1890), v.
2. *Human Body*, 532. No name was given for the newspaper. The city was Newark, NJ and the date was May 20, 1876.
3. *Human Body*, 398. *Republican* (Omaha, NE), Jan. 15, 1874.
4. *Human Body*, 411. *Gazette* (Davenport, IA), Feb. 1, 1874.

her earlier advocacy of free love and her later eugenics. In a very real sense the one replaced the other at the top of her social agenda.[5]

Slavery became dependency in what a Nebraska newspaper would refer to as "the social question... having special reference to the dependent classes, the women, children, maimed, insane and idiotic."[6] In modern terms, Woodhull's scheme for "raising society" was eugenics taken to include not just breeding for a superior body and mind, but dealing with crime, prostitution, marriage and child-rearing. Give women the education to think eugenically, she said, and the power to act eugenically and our social problems would disappear. That's precisely what she told an audience in Dubuque, Iowa.

"The great demand of the age is for better men and women. But here comes a woman [Woodhull herself], ready to tell you out of the fulness of a mother's heart how to bring into this world better men and women, and you start back with horror!... A great deal has been said about the prevention and cure of prostitution; but little or nothing has been done toward accomplishing it. I will tell you how to accomplish it: send your daughters out into the world as peers of your sons; teach them that it is honorable for women to earn their living—and then give them a chance to do so." Proceeding to describe what marriage *should* be, she claimed that to bear a child is the most sacred and honourable mission on earth. The pregnant woman is a co-worker with God in giving to the world an immortal being.... Preachers turn all their attention to saving souls when they would be in much better business saving bodies.[7]

On the surface, Woodhull's plan seemed to empower women just as she claimed. They are to be allowed careers, so the poor did not need to become prostitutes and the middle-class and above were not forced to marry for money. Marrying for love, their attitude of acceptance for the child as it grows in the womb will, she believed with all her heart, produce superior children.

But is that how things work? As many career women today will attest, a successful professional career does not automatically mean a happy marriage or joyful motherhood. If failing in either of those areas means a child with criminal inclinations, then women have exchanged dependency for a far worse burden. One reporter explained Woodhull's view of crime by quoting remarks in which she hinted that when a child murders, the mother bears the guilt.

"There is one hope still, which is that men and women will meet in solemn conclave and discuss the purity of the social question. If it was rightly understood, the prisons would not be filled. I have asked mothers with bad children if they wanted these children, and they would answer, 'No, Mrs.

---

5. For more on her early ideas, see *Free Lover: Sex, Marriage and Eugenics in the Early Speeches and Writings of Victoria Woodhull* (Seattle: Inkling Books, 2005).
6. *Human Body*, 397. *State Journal* (Lincoln, NE), Jan. 13, 1874.
7. *Human Body*, 415–16. *Times* (Dubuque, IA), Feb. 3, 1874.

Woodhull; I tried to murder them unborn.' There is no blushing. The men and women who stand pure before the people need not blush. Murder is stamped on more than one woman for neglecting to inform their children of their natural condition."[8]

Two weeks later, speaking in Iowa, she hinted again that the answer to the social question lay in the underlying attitude women have toward their role bearing "the image of God in reality." Women, it seems, were to blame for our social ills because they were not taking their responsibilities as mothers seriously enough.

> ... She said that the idea of becoming a mother is something woman has never looked upon with enough sanctity; and just as soon as the mothers of the country commence to think, prostitution will cease. All she ever asked was that the basis of marriage should be *love*—educate your daughters to marry for love and not for money or homes.[9]

What Woodhull meant came through in even more concrete terms two days later in an Iowa paper when she attacked mothers who, "try to murder their children before birth and then wonder why those children when grown to be men, turn out murders."[10]

Contemplating abortion wasn't the only way a mother could turn her unborn child to evil. That same month in Minnesota she mocked the prayers of temperance advocates. (Bolding added.)

> "What an idea this is of your women going round and praying at saloons to the keepers. If the mothers would make no more drunkards there would be no more trouble. When I was at Clinton Junction on my way here, I was stopped by a man who said he wanted to speak to me, and hoped I would not be offended. He exposed his breast to me, which was marked with a bottle. He said, 'When my mother was carrying me she went into a saloon and was seized with an unconquerable desire to drink.' She brought forth a drunkard. **Let the mothers agree to breed no more drunkards and there will be no use for saloons. It is the women who make these drunkards.**

> "The day must come when the study of the laws and relations of the sexes be made a pure and holy thing if we would have better men and women. If a man stocking a farm should act with so little foresight and discretion as men and women do in making children, he would be called a fool."[11]

Far from freeing women, Woodhull's arguments put them in a far worse bind. Unless a potential mother conformed to these "laws" about sex and made

---

8. *Human Body*, 397. *State Journal* (Lincoln, NE), Jan. 13, 1874. Informing children "of their natural condition" may have meant warning that their mother's behavior while they were in the womb had stamped them with certain evil but 'natural' inclinations.
9. *Human Body*, 410. *Gazette* (Davenport, IA), Feb. 1, 1874.
10. *Human Body*, 415. *Times* (Dubuque, IA), Feb. 3, 1874.
11. *Human Body*, 419. *Pioneer* (St. Paul, MN), Feb. 13, 1874.

them "a pure and holy thing," she would be responsible not just for her own crimes, but for those committed by her children. Even before it developed an emphasis on genetics, eugenics was brutally coercive.

About this time, Woodhull may have sensed that eugenics based on folk ideas about how a mother's experiences and feelings affected her unborn child were weak, at least with some audiences. Her February 1874 reference (above) to "the laws and relations of the sexes" and "a man stocking a farm" suggest that she felt a need buttress her speeches with more science, while the mention of "a pure and holy thing" suggested she also felt a need to tie her lectures more closely to the existing, Bible-based religiosity.

In the newspaper articles she published as "Press Notices" in *Human Body*, the first indication she was trying to find her ideas in the Bible in a systematic way came in August of 1875 from Albany, New York. (Bolding added.)

> Mrs. Woodhull lectured at Martin's Opera House, last evening, upon Bible mysteries. She based her reasonings upon some texts of the Scripture, which she read from the prophecies of Daniel, the Revelation of St. John, and the history of Creation. **She assumed that the Garden of Eden was intended to mean the human body, and that the temple of God was the same thing.**[12]

The very next day she spoke at another opera house in Troy, New York. There, she based her arguments on science and Charles Darwin's then sixteen-year-old theory of evolution. A reporter had nothing but praise for her "eloquence" and "fervour." The excitement Woodhull could induce in some reporters was amazing. (Bolding added.)

> ... It is certain that Mrs. Woodhull utters truths and advanced ideas worthy of consideration. She considered the American people the culmination of the development of nations, and paid a high tribute to their superiority over other races. The fearless and earnest manner in which she discussed the sexual question enchained the closest attention. Mrs. Woodhull presented a strong argument to prove that the triumph of her doctrines would do away with prostitution and crime, empty our jails and penitentiaries, and introduce a millennial era, insuring a lofty plane of moral, mental, and physical development. **She believed that the process of evolution would eventually produce a perfect woman, possessing every virtue and worthy attribute of her sex, from whose progeny would spring a perfect race.**
>
> In an impassioned outburst of eloquence, she charged the responsibility for the evils of intemperance and prostitution upon the mothers of our race, saying it was their duty to instruct their children in matters appertaining to their bodies, and not leave it for others to do.[13]

---

12. *Human Body*, 441. *Argus* (Albany, NY), Aug. 21, 1875.
13. *Human Body*, 441–42. *Sunday Trojan* (Troy, NY), Aug. 22, 1875.

Woodhull seems to have spoken on both topics under the title, "The Human Body the Temple of God." If she spoke for two nights, she might give both lectures. At least that's how the *Morning Whig* (Troy, NY), quoted below described her two speeches there. And if she spoke for one night, it made sense to choose the one she thought most appealed to her audience—in the United States perhaps the religious argument.

In 1875, Woodhull published the religious argument as a booklet called "The Garden of Eden; or, Paradise Lost and Found." Years later (1890), when she published *The Human Body the Temple of God*, that booklet became the book's first chapter and here it is included as Chapter 4. There you will find a strained and heavily allegorical argument, "that the Garden of Eden was intended to mean the human body, and that the temple of God was the same thing." Finding secret messages in religious texts has much the same appeal (and makes as little sense) as conspiracy theories in politics.

That raises the question why she didn't include her scientific and economic arguments for eugenics in her 1890 book. There were probably two reasons. First, as she noted in the preface, she wanted the book to include what she had lectured on "from 1869 to 1877… throughout the length and breadth of America." Including topics such as "The Argument for Woman's Electoral Rights," along with over 300 pages of "Press Notices," pushed the book to over 600 pages, leaving no room for more on scientific eugenics as she understood it. Second, she had already published one booklet with scientific arguments for eugenics, *Stirpiculture*, in 1888, and published one on the economic and political issues related to eugenics that same year, *Humanitarian Government* (1890). The following year she would published what may be her most widely circulated booklet on the topic, *The Rapid Multiplication of the Unfit* (1891), and in 1893 she would publish *The Scientific Propagation of the Human Race*, which she said was based on what she had lectured on "throughout America, from 1870 to 1876." It is easy to suspect that she left her non-religious arguments for eugenics out of *Human Body* because she wanted to publish them as booklets that would be more widely read. Finally, publishing the book in London, it made good marketing sense to use the title of her best-known English speaking series as the title, even though much of the book's content had little to do with the speeches she made under that title.

Over time, the economic and scientific arguments for eugenics seem to have become more important for Woodhull than the folk tradition and religious ones. When her October 31, 1883 marriage into the wealthy Martin family gave her enough money to self-publish, she chose to focus on those arguments.

We also need to be careful that we don't interpret Woodhull's claims about her speech topics too narrowly. Woodhull was an excellent speaker and rarely read from notes, particularly after her introductory words, so it's likely that she adjusted what she said to her audience and continually improved it to broaden its appeal. That partly explains the praise she received from many newspaper reporters. She was a excellent and persuasive speaker.

One reporter, however, was more objective than most in describing two nights of lectures in Troy, New York on Saturday and Sunday, August 21–22, 1875. The Saturday night lecture follows the general theme of *The Scientific Propagation of the Human Race* and even uses that term. The reporter remarked on "her well-known theory that a perfect race can consist only of perfectly formed men and women."[14] That suggest that in 1875, eugenic ideas were not only becoming "well-known," they were attached in the public mind almost exclusively to Woodhull—they were "*her* well-known theory." No mention is made of Francis Galton, but that's hardly surprising. After all, Galton was not traveling around the United States, speaking almost nightly to large audiences in cities and towns across the land. In fact, by his own admission, he was doing little to promote eugenics. Notice too that in 1875, her speech included folk beliefs about a mother's influence on the child in her womb, as well as "inherited characteristics from their parents." (Bolding added.)

> She next showed that the misery, vice, and crime, with which the world is cursed, exist because the propagation of the race is carried on without any regard to the results obtained: **that the criminal and vicious classes were made so by their mothers during gestation, or by inherited characteristics from their parents.**... The criminal classes recruited constantly from the children born of mothers who did not want them—in other words, from unwilling or undesired children. **To rid the world of all these classes it only requires to place women in such a position that they will never bear children except when they want them, and this is her whole right.... Every impression and thought, and especially every strong desire of the mother, has its effect upon her unborn child.** Mothers do not realize this, however, and her opposers do not intend that she shall get their ear to awaken them to their terrible responsibilities in this regard.[15]

The remark that "mothers do not realize this," brings up another theme Woodhull repeated often. Social ills would only end when mothers gave their children, and particular their daughters, the proper sex education, one that included eugenics ("the process of evolution"). This idea existed at least as early as her 1871 "Children—Their Rights and Privileges" speech (Chapter 2), and that's precisely what she would tell an 1875 Vermont audience.

> "... When the young are taught the laws of life and taken into the heart and confidence of parents the world will become more virtuous. No one has loose ideas when discussing the ways of improving stock. In the fine art galleries of the old country where are statues of men and women true to nature, ladies and gentlemen pass along without blushing, because there is nothing

---

14. *Human Body*, 442. *Morning Whig* (Troy, NY), Aug. 23, 1875
15. *Human Body*, 443. *Morning Whig* (Troy, NY), Aug. 23, 1875.

to blush for. The people who dare not discuss these subjects are the vulgar, the impure, the ignorant, and the vile. **All this vulgarity, impurity, ignorance, and vice must be eliminated by the process of evolution, through discussion, virtuous habits, education and intelligence."**[16]

Woodhull would echo that theme fifty-two years later when she told an AP reporter "I advocated that fifty years ago in my book, *Marriage of the Unfit*. I am also glad that parents are now beginning to instruct their adolescent children in the facts of life."[17] She doesn't explain why children would heed difficult to understand (much less obey) eugenic teachings about the "laws of life," when they routinely ignore easily understood lessons about right and wrong. Woodhull's critics seemed to have focused on condemning her teachings as vulgar, as she so clearly hoped. They would have been wiser to criticize her scheme as impractical and ineffective.

Returning again to the detailed description of her Troy, New York speeches five weeks earlier, we find the reporter describing something closely resembling her published "The Garden of Eden." (Bolding added.)

> Last evening's lecture was an exposition of Mrs. Woodhull's understanding of revealed religion. She spoke from the texts: Daniel xiii, 8 and 9, and Revelation x. 7.
>
> The Bible, Mrs. Woodhull thinks, is a book which is sealed to ordinary mortals. The curse that is put upon woman is: "Thy desire shall be to thy husband; and he shall rule over thee." All the sin in the world comes as a result of this curse, and the curse itself came as a result of the violation of the law of nature.... Through education and knowledge on this subject she expects the strong carnal propensities to grow less and less, mankind to be physically improved and woman especially, and the perils of maternity to be removed, because the pollution of the body will be at an end.
>
> The human body she regards as the Temple of God, and the fall of Eve the subjection of woman to man. **The garden of Eden is simply the human body. All the words relating to salvation refer to the salvation of the body. "Keep my saying, and ye shall never die," refers to bodily life.** The beginning and end of the Bible establish its unity, the first part speaking of the tree of life from which if Adam and Eve were not shut out they would eat and live for ever; and the last, of the tree of life, "the leaves of which were for the healing of nations, and from which we are to get a pure river of water of life and die no more." **The perfection of living that the speaker wishes to see is to purify the stream of life; and the woman, once the slave of man, is to be redeemer of the race.**[18]

---

16. *Human Body*, 454. *Union* (Kenosha, WI), Sept. 30, 1875.
17. "Says Voting at 25 is 'Young Enough.'" *New York Times* (May 3, 1927), 6. Quoted in *Pivot*, 31.
18. *Human Body*, 443–44. *Morning Whig* (Troy, NY), Aug. 23, 1875.

In the fall of 1876, Woodhull was able to return to Boston, after having been unwelcome in the city since exposing the unsavory Beecher-Tilton affair four years earlier in 1872. What she said to Bostonians is interesting. In some ways, she was ahead of her time in championing fashionable ideas, particularly in her beliefs that problems connected with sex flow from an unwillingness to talk about it and that openness is a kind of moral antiseptic. "The time is coming," she told the city's bluestocking matrons to loud applause, "when everyone of any intelligence will see that there is nothing vulgar save ignorance."[19]

On the other hand, Woodhull was also capable of the same excessive of sentimentality for which the Victorian Age is famous, although it is not clear whether she believed what she was saying or simply said it because her audience would approve. Here's one example.

> She then drew another picture of the child's receiving the information it sought from a holy woman, a pure mother. Who made you darling? Mamma carried you under her heart days, weeks, and weary months, and at last went into the Garden of Gethsamane to bear you into the world. Now, my precious child, you can see why mamma loves you so; why she would give her life to save yours and basing its whole after-life and the current of its thoughts and actions on that frank avowal: she said, in conclusion, that child would never commit an act of which it would not dare to tell its mother, because its mother had revealed concealment unnecessary and out of the question.[20]

Although her Boston audience may have approved, the idea made little sense. Sons and daughters of such a mother would be at least as likely to conceal actions from her because they have no desire to change their behavior, but see no reason to hurt her with a confession. Those who thought otherwise over-estimated the power of mother-induced guilt.

You can also see in Woodhull's Boston speech an early example of a debate that haunts eugenics to this day. Eugenic *activists* insist we know enough to act now to create a much better world. Eugenic *moderates*, while agreeing on the ultimate goal, insist that we don't yet know enough to act, except perhaps in a few limited areas. Woodhull was clearly an activist, perhaps the first outspoken activist for modern, scientific eugenics. She believed the main problem lay in the easily remedied ignorance of mothers about facts they should know.

> ... "I ask of every mother never to bear a child that can by any possibility fill a criminal cell or an idiot room. I ask that our mothers understand in all its importance this mighty problem: I ask that the ignorance which now hides it from her vision be at once and for ever dissipated, even though it exposes the truth in all its horrible and ghastly realism. You patronize horse trots and cattle shows; you discuss publicly, and have it reported in

---

19. *Human Body*, 549. *Boston Herald* (Boston, MA) Oct. 2, 1876.
20. *Human Body*, 551. *Boston Herald* (Boston, MA) Oct. 2, 1876.

the newspapers, how to raise Durham bulls, and how to create fine stallions, and how to graft the good elements of one animal into those of another, and nobody remarks it; but if the poor mother, torn by conflicting emotions, racked with an agony none but a mother can conceive or realize, cries out in despair. 'In the name of God, tell me how to create my child: tell me, in order that I shall not bear an idiot or a criminal," everyone would hold up their hands in holy horror. 'Oh! She's vulgar: don't go near her,' they would say."[21]

By "truth in all its horrible and ghastly realism," Woodhull probably meant her folk eugenics. Linking a mother who contemplated abortion to the son who later became murderer or a visit to a bar to an alcoholic daughter is "horrible and ghastly," although it would provide a warning to mothers-to-be how "not to bear an idiot or a criminal."

But as eugenic moderates later pointed out, it's difficult to go from the mating of "Durham bulls" to human society. We know precisely what we want in cattle and a bull who delivers that can be mated with hundreds of cows, with his service advertised in Boston's most high-toned newspapers, as Woodhull seemed to suggest. Are we to have similar human 'stud farms,' where the women, having heard of the service through a local newspaper, line up for their few minutes with a man from superior stock? And when a bull is born that doesn't measure up, he is castrated and slaughtered a short time later for meat. It's hard to come up with a human equivalent for that. Finally, all the arguments for men apply with equal force to women with one exception. Because they must carry the child, women have far fewer potential offspring than men. This means a superior woman is a scarcer resource. We might exempt a particular man from doing his eugenic duty because another could work a bit harder in his place. But it's much harder to tell Gertrude that she must now bear ten children, one every two years of her life from twenty to forty, because Helga refuses to bear her assigned five.

Woodhull seemed to realize that some coercion would be necessary to achieve the results she wanted. Later that evening she told that Boston audience, "Tonight, if I had the power, I would make it impossible. I would make it a crime for men and women to marry ignorant of parental responsibility. They have no right to marry and people these abominable institutions, for it is almost entirely from such sources that the recruits to these places come from."[22]

It is true that she went on to hinge her ban more on *not knowing* about eugenics rather than for being of inherently bad stock. Continuing the remarks above and to the great applause of her audience, she tells of a son, on being told the eugenic 'facts of life,' responding with, "Mother, dare I marry?" and a daughter with, "Mother, I do not know that I am worthy of marriage." But

---

21. *Human Body*, 550. *Boston Herald*, Oct. 2, 1876. Forty years later, Margaret Sanger would use an almost identical 'women want to know' argument for birth control, although her eugenic agenda would be better concealed.
22. *Human Body*, 552. *Boston Herald*, Oct. 2, 1876.

that's hardly realistic. Children that responsible will make excellent parents. It's the ones who, on being told "You're a drunk and any children you have will end up drunks," that could care less who are society's greatest worry. Forcing potential parents to hear about eugenic laws would quickly evolve to forcing them to abide by those laws, willing or not.

It also appears that Woodhull was now no longer a lone voice calling for a more eugenic motherhood. In 1876, newspapers were reporting on scientific research that claimed to demonstrate the same link between bad mothers and crime that Woodhull had been speaking about at least since the spiritualism convention of 1871. In her Boston speech, Woodhull brought up those reports and used them to justify her own controversial speeches: "As long as the mothers of America read the statement in the daily papers unblushingly that one thousand criminals had descended from one Margaret, surely I have nothing to fear from the discussion of this question."[23]

The "Margaret" to which she referred was Margaret Juke, a woman that Richard L. Dugdale would brand as "Margaret, the Mother of Criminals" in his book, *The Jukes: A Study in Crime, Pauperism, Disease and Heredity*, which would be published the next year (1877). Dugdale kept his raw data secret, but we now know that her real name was Margaret Robinson Sloughter.

Dugdale began his research in the summer of 1874, when he visited 13 county jails in New York state, asking prisoners questions about their heredity and environment. He found no overlapping family ties until he reached Ulster county, where he found six inmates under four family names who were related. Further research in the community led him to 709 people who shared a common descent by blood or marriage, a number he claimed would have grown to 1200 if he could have traced all the descendants of six original sisters. Among those he could trace, he found 180 who had received public assistance, 140 criminals, and 50 prostitutes. He estimated the family had cost society over 1.3 million dollars between 1800 and 1875. The following year (1875) his research was made public when it was published in the annual report of the Prison Association of New York.

Dugdale's research provided added scientific respectability for Woodhull and may have caused her to increase her stress on science, economics, and negative eugenics. A generation later, Margaret Sanger's new birth control movement would receive a similar boost from Henry Goddard's 1912 study of a New Jersey family in *The Kallikak Family: A Study in the Heredity of Feeble-mindedness*, as well as Arthur H. Estabrook's updating of Dugdale's research in *The Jukes in 1915*, published in 1916. In the eighth chapter of her 1922 *The Pivot of Civilization*, Sanger would note: "Eugenics is chiefly valuable in its negative aspects. It is 'negative Eugenics' that has studied the histories of such families as the Jukeses and the Kallikaks, that has pointed out the network of imbecil-

---

23. *Human Body*, 553. *Boston Herald* (Boston, MA) Oct. 2, 1876.

ity and feeble-mindedness that has been sedulously spread through all strata of society."[24]

Today, that early eugenic research is regarded with suspicion. Investigations that have broken through the veil of secrecy created by concealing the actual family names have found that some members of those families were successful and well-respected members of their communities, discrediting claims that the problems of the others were inherited rather than environmental.[25] And the difficulties of other members of those families can be explained as easily by environment as by heredity.

Unfortunately, in the latter half of the 1870s eugenic ideas and the notions of progress they inspired were becoming more widely accepted, particularly in the press, where an awe of science isn't always accompanied by an understanding of what careful science involves. As a result, the shift in the emphasis of Woodhull's speeches from free love to eugenics, won her the praise of newspaper editors, as noted in this 1877 editorial in the *Detroit Post*. That editorial, written in advance of her speech, was little more than a free advertisement for all "thinking people" to hear her speak. (Bolding added.)

> Mrs. Woodhull will deliver, at Whitney's Opera House to-morrow evening her famous lecture, "The Human Body the Temple of God." This lecture has been delivered by Mrs. Woodhull in all the large cities of the east to immense audiences composed of the thinking people of each community. In Boston her audience numbered over four thousand persons. **A decided change has come over the papers of the country in the last two years in their treatment of this remarkable woman.** It was natural that the subjects she discussed should startle people; and the shock manifested itself in the press by tirades of abuse, which was more indecent even than the sentiments and utterances they falsely ascribed to their object. **Since then, however, thousands in all parts of the country have heard Mrs. Woodhull for themselves, and have learned for themselves that she discusses questions of vital importance to the happiness of the human race with as much purity as eloquence of expression and with strong, if not invulnerable, logic.** Whether people sympathize or not with her remedies, they have ceased to deny the existence of the disease, and the press, taking its tone from the change, now announces her advent and criticizes her utterances at least in respectful language.[26]

---

24. Chapter XIII of *The Pivot of Civilization in Historical Perspective* has excerpts from *The Kallikak Family*.
25. See Scott Christianson, "Bad Seed or Bad Science: The Story of the Notorious Jukes Family," *New York Times* (Feb. 8, 2003).
26. *Human Body*, 567. *Detroit Post*, July 2, 1877.

Woodhull in the U.K.

Shortly after that speech, Woodhull moved to England, where she would continue a eugenic campaign that would earn her increasingly respect, particularly among certain influential groups of women. Judging by the articles she published, her first speech seems to have been in Nottingham on Tuesday, September 4, 1877. It was covered by at least three local newspapers.

One paper noted that her audience was "highly respectable, many of our local dignitaries being present, principally ladies."[27] It was a pattern that would repeat with Charlotte Perkins Gilman and Margaret Sanger. Affluent women seemed to find eugenics particularly attractive, particularly when their thoughts turned to the plight of poorer women who, they thought, were having too many children. That same paper noted that Woodhull began with a biblical quotation, "Know ye not that ye are the temple of God...," so the speech seems to have been a version of "The Garden of Eden" which lasted two hours, according to another newspaper. The second newspaper also mentioned the emphasis Woodhull placed on mothers awakening in their children "the responsibilities of life and especially maternity."[28]

Though she liked causing a fuss, Woodhull also wanted her ideas to be respectable. Eugenics taught from mother to daughter was perhaps the least controversial approach she could take, although it was also the least effective. It is unlikely that 'unfit' mothers among the Jukes and the Kallikaks were going to tell their daughters, "I'm so awful, I shouldn't have had you as a daughter, and you're as bad as me, so you shouldn't have any children." Negative eugenics is always most popular with those who think it does not apply to them.

Two weeks later, on Thursday, September 27, Woodhull would be in Liverpool, again speaking on "The Human Body the Temple of God" before what the *Liverpool Post* would call "an audience of high standing in life"[29] And while in the U.S. her eugenic ideas had become well-known, a Liverpool reporter would find them "novel," suggesting that Woodhull was right when she claimed to be the first to popularize eugenics in the U.K. (Bolding added.)

> ... She is certainly a splendid actress, and an orator of rare power. **She has taken up a novel line.** Her burden was that if all the mothers in the world would rise to a sense of their responsibilities, and act accordingly, the condition of human nature would immediately undergo a vast improvement.[30]

Two weeks later, on Monday, October 15, she spoke in Manchester, and thirteen years later she still retained an adoring letter from someone who heard her speech. It claimed that if every mother were like her, "we might pull down nine-tenths of our prisons," and closed with "May you conquer here as

---

27. *Human Body*, 580. *Nottingham Daily Express*, Sept. 5, 1877.
28. *Human Body*, 581. *Nottingham Guardian*, Sept. 5, 1877.
29. *Human Body*, 581. *Liverpool Post*, Sept. 28, 1877.
30. *Human Body*, 582. *Liverpool Albion*, Sept. 28, 1877.

you have conquered in America.³¹ For good or ill, her eugenic message was now being repeated on both sides of the Atlantic.

There were other letters and articles. In a letter about a Tuesday, December 11, 1877 speech, a Westminister pastor wrote that her cause deserved the "moral support of the country" and promised to spread her message to his "own audiences."³² Jumping ahead over two years, an 1880 article in *Christian Union* demonstrated that the religious press could be as outlandish in their praise of her odd explanation of "the hidden mysteries of the sacred writings,"³³ as the secular press was about her use of science.

The final article in Woodhull's collection came two and a half years later (1882) in a London newspaper. It offers a good illustration of the enthusiasm some had for her ideas. (Bolding added.)

> ... So long as marriage is simply the expedient to provide a shelter and food and necessaries of life so long will man (God's image) continue to create murderers, idiots, drunkards, thieves, and scoundrels, and prisons flourish, mad-houses be built, asylums needed, Marwood paid, and gallows erected: but once let the holy teachings of Mrs. Woodhull be understood, and marriage will be the thing of choice, not of necessity....
>
> **Mrs. Woodhull would have none enter upon the holiest of all relations—marriage—save in a devout spirit; the divorces would cease, suffering would pass away, beauty, health, and enjoyment would be everywhere, and sad sights of suffering humanity would no longer meet our view.**³⁴

Looked at from over a century later, it's difficult to know how to respond. A "devout spirit" about marriage and a vague mysticism about motherhood (later called the 'feminine mystique') would do little to prevent the birth of murderers, thieves and the rest. To the extent that parents influence their children, good children come from good parenting rather than from the ideas of a speaker who appealed most to those idle rich women who leave their children to the rearing of nannies. And in an age when little was known about genetics, the best chance for healthy, honest children lay in ignoring Woodhull's advice, being practical, and marrying a man who was himself healthy, honest and capable enough to provide "shelter and food and necessities of life."

On the next page is the May 8, 1927 (pt. 2, p. 6) *New York Times* article about Woodhull and forced sterilization that was discussed in Chapter 1. It is perhaps the last major press notice she received during her well-publicized life.

---

31. *Human Body*, 583–84. E. E. M. letter of Oct. 16, 1877.
32. *Human Body*, 584. Rev. B. letter of Dec. 12, 1877.
33. *Human Body*, 585. *Christian Union* (London), Feb. 13, 1880.
34. *Human Body*, 586–87. *Advertiser* (London), Aug. 12, 1882. "Marwood paid" refers to London's well-known hangman, William Marwood (1820–83). During his nine years in the office, he was paid £10 for each of 176 executions.

SAYS VOTING AT 25 IS 'YOUNG ENOUGH'

Mrs. Woodhull Martin Supports the British Franchise Bill to That Effect.

BRIGHTON, England, May 7 (AP).—Mrs. Victoria Claflin Woodhull Martin, the first woman candidate for the Presidency of the United States in 1872 and long a fighter for equal suffrage, believes that 25 is plenty young enough for men and women to obtain the franchise.

In 1872 Mrs. Martin carried the banner of equal suffrage in Maine and California as Presidential candidate of the Equal Rights Party and at 88 she is still interested in promoting the emancipation of women.

"I want women to have the vote as soon as they are fit to use it," Mrs. Martin told a correspondent for The Associated Press, "but I do not believe in forced maturity. Twenty-five is young enough for persons of both sexes to exercise the franchise."

Mrs. Martin was seated with her daughter, Miss Virginia Woodhull, in their apartment here when she received the correspondent. Time has not dimmed the eyes of this spirited woman who, with her sister, the late Lady Cook, formerly Tennessee Claflin, was the first woman broker in New York and lectured and published Claflin's Weekly in support of equal suffrage and eugenics before they both came to England.

The surprised interviewer, who expected to find the advocate of equal suffrage ready to defend the "flapper vote," as it is termed by opponents of the bill to give British women the vote at the age of 21, the same as men, instead of 30, asked:

"But what of yourself at 21?"

"I was making history when I was 21," Mrs. Martin replied. "But I was a wife when little more than a child. My son was born when I was very young and I had an unusually advanced education at home. My case was exceptional."

"But what about your daughter?" Mrs. Martin was asked, "surely she was fit to vote at 21."

"Certainly not," she replied.

"Mother is right," Miss Woodhull agreed with good humor. "I knew nothing when I was 21, although I was studious and had read a great deal. I question if even the modern emancipated girl is a sufficient judge of character to discriminate between political candidates."

Mrs. Martin, who wrote and lectured for thirty years on eugenics, remarked that she was pleased to read that the Virginia Eugenics law had succeeded in establishing the right to sterilize the feeble-minded.

"I advocated that fifty years ago in my book, 'Marriage of the Unfit,'" she said. "I am also glad that parents are now beginning to instruct their adolescent children in the facts of life. My sister, Tennessee, and I were mercilessly slandered fifty years ago when we dared advocate women's emancipation and discussed eugenics in America, but time has proved that we were right."

CHAPTER 4

The Garden of Eden

Introduced by Michael W. Perry

> *With a perfect physical body—man reconciled to God—all other perfections follow as its fruit, necessarily. The opposite proposition to this is the stumbling-block over which all Christians have fallen; they have given all their attention to saving the soul hereafter, when this salvation depends entirely upon saving the body here and now.*

Victoria Woodhull enjoyed portraying herself as a martyr for truth. While most of her opponents seemed to have been the naive, traditional, and religious sort who played into her hands by opposing her in ways she chose for them, there is an interesting example of a more snobbish criticism. It came a few months after Woodhull moved to England. In a byline dated December 8, 1877—indicating the speech came shortly before that—the chief London correspondent for *New York Times* described a lecture by New York City's former celebrity. At the end of Chapter 3, we looked at newspaper coverage of Woodhull's earlier speeches in the U.K. (Nottingham, Liverpool, and Manchester). Here we explore what may have been her first London speech, as well as the first in a series of lectures she would give at St. James Hall over several years. This particular one seems to have been promoted by "hoardings"—signs placed on fences and walls of buildings. Later she would use newspaper advertising.

The *New York Times* article is on the next page, with the unrelated portions digitally removed before and after the remarks about Woodhull. In it, you will discover that the Prince Malcolm Khan, mentioned in the *Evening Standard* article (Chapter 1) claiming that Woodhull was "twenty-five years ahead of her time," was the owner of "Khan's Museum of anatomical figures." He and Woodhull were in similar businesses titillating the public with anatomy.

According the article, Woodhull's lecture was on "The Human Body the Temple of God." In her 1912 interview in the *Evening Standard*, Woodhull claimed that the 1877 speech with which she has "amazed London" had been "The Scientific Propagation of the Human Race." Which was it? As we pointed out in Chapter 3, when her lectures ran for two nights, her "Human Body" series may have included one night on a religious theme, "The Garden of Eden" (this chapter) and one night on "scientific propagation" similar to Chapter 8 in this book. Since she could speak well without notes, on a single-night lecture, she may have blended the two lectures together, changing the emphasis to suit the audience. Also, since a third of a century had passed, she may have gotten the dates confused. As we note later, she did give a speech on scientific propagation in 1893 at St. James Hall. There might have been an earlier one, not advertised in the *Times* (London) but also not her first speech in 1877.

CURRENT ENGLISH TOPICS.

AMERICAN GENIUS UNRECOGNIZED.
A DISTINGUISHED TRAGEDIAN CRUSHED IN LONDON — VICTORIA WOODHULL'S ARRIVAL AT THE BRITISH METROPOLIS — AN INDECENT LECTURE — STRANGE ENGLISH SOCIETY — AGGRESSIVE RITUALISTS.

From Our Own Correspondent.
LONDON, Saturday, Dec. 8, 1877.

The other morning I was attracted by the latest proclamation of transatlantic genius: "Victoria Claflin Woodhull, the great American orator, will deliver her sparkling lecture, 'The Human Body the Temple of God,' at St. James' Hall. Letters to be addressed care of the manager." On the hoardings of Willing, the billsticker *par excellence* of the metropolis, was posted attractive portraits of Victoria Claflin Woodhull. Her effigy appeared in the shop windows, with some religious legend printed underneath. She attracted a crowd, and possibly may go on doing so for some little time, because she is the great American orator, and people want to see what a great American orator is like, especially when the speaker is a woman, and more especially when she is supposed to speak publicly about subjects which never even enter the thoughts of decent and respectable women. American ladies are supposed to speak with greater freedom and to enjoy a wider margin of liberty than English ladies; and Victoria Woodhull is regarded by some people as the outcome of this greater freedom, the product of a more liberal education and a freer intercourse between men and women in the discussion of physiological and kindred subjects. But I feel sure we shall find Victoria out, and American papers will tell us that she is not "the great American orator," while traveled English men and women know that the women of the United States are not in the habit of talking about the human body in miscellaneous company in a way that would bring the blush to the cheek of her average British sister. Mrs. Woodhull will come to grief between the Scylla of prurient curiosity and the Charybdis of current philosophy. Those who went to St. James' Hall to hear something nasty, came away disappointed, because she didn't go far enough; those who went to learn something discovered that Mrs. Woodhull had nothing new to tell them. It was a distinguished audience, and at the same time a curious one—Journalists, critics, essayists, students of philosophy, men about town, strong-minded women, precocious boys and girls. No wonder the orator was nervous. She must have felt that she was not in the presence of an ordinary crowd. They treated her with the greatest possible respect; they did not even titter at suggestions, which, coming from a woman, might have excited a spirit of badinage in a less cultivated assemblage. You know all about the lady and her views. It is therefore unnecessary for me to discuss them. The point upon which her discourse turned was the faulty education of boys and girls in questions which are not discussed in general society. Mrs. Woodhull wants English mothers to tell their children all she knows, to give them the benefit of her experience, and to win the confidence of their children sufficiently to have them tell her their prurient secrets, wishes, and thoughts, with a view to directing their passions. That plainly is what the orator told her hearers; and it is upon such an extension of our educational system that the future greatness and happiness of the English people depends. "Letters to be addressed care of manager" is, I presume, an invitation to ladies and gentlemen, to mothers and fathers, to boys and girls, to write to her, with a view to making converts. Mrs. Woodhull will be sure to receive many letters, and to have many callers. I venture to predict that her correspondents will not be philosophers, nor her callers respectable women. There are a set of weak-backed sneaks in London who devote their wretched lives to running down women, to Argyle orgies, to the study of indecent pictures. They will be curious about Mrs. Woodhull. She is rather a good-looking woman, and she has circulated her picture. Fast men will think her great fun. *Roués* in search of a new sensation will call upon her in the disguise of philosophers. Over-educated young women who study anatomy and want to be doctors will be among her visitors. She will attract just the sort of people who used to go to Khan's Museum of anatomical figures. The wise and intellectual people who went to hear her on the first night will go no more; she has no news for them. The flippant, the morbid, the unhealthy people who come under the special care of the Society for the Suppression of Vice, will be Mrs. Woodhull's chief patrons. The philosophers have already found her out. Her appearance, her manner, her half-nervous style of utterance, her little womanly ways, so out of keeping with the matter of her lecture, pleased her audience from an artistic point of view, though she has neither the voice, the *chic*, the eloquence of Kate Field, whose oration on Dickens is spoken of on all hands as a poetic and intellectual effort. The English prejudice against women on the platform has died out—at all events in London. Kate Field might make a successful tour. But it is to be hoped that this growing liberalism of sentiment in regard to what women may and may not do will never extend to the patronage of female lecturers on anatomy, ethics, and libidinous philosophy.

Judging by the remarks in the *New York Times*, the subject of her first London lecture was the human body discussed in ways not usually employed at that time in mixed company. Nevertheless, the writer felt that the audience, obviously not an "ordinary crowd," came away disappointed that they did not hear "something nasty." Some of what she did say concerned what is today called sex education: "Mrs. Woodhull wants English mothers to tell their children all she knows, to give them the benefit of her experience, and to win the confidence of their children sufficiently to have them tell her their prurient secrets, wishes and thoughts, with a view to directing their passions."

We get a glimpse into that lecture through an 1890 book bearing the same title as the lecture, *The Human Body the Temple of God* (1890). As she noted in the Preface, the book collected under on cover what she wanted remembered about her speeches "from 1869 to 1877… throughout the length and breadth of America." The first chapter, a republication of her 1875 "The Garden of Eden; or The Paradise Lost and Found," suggests what she might have said in her late 1877 speech in London's St. James Hall, keeping in mind that its length, 56 printed pages and roughly 22,000 words, was too long for a single evening's speech. Given the need to speak slowly and clearly in an age before sound amplification, the whole might have taken over three hours to deliver.

That night Woodhull may have cut out the first half of "The Garden of Eden," which contains her theory that the early chapters of the biblical Genesis were highly allegorical. As she put on page 29 of the original: "Suppose, I say again, it should after all turn out that the long-lost Garden of Eden is the human body; that these three, the Kingdom of God, the Temple of God, and the Garden of Eden, are synonymous terms and mean the same thing—are the human body." That, she went on to explain, meant that we should reject the idea that, "any of the parts of the body can be vulgar and impure, and unfit to be discussed either in the public press or the public rostrum?" Of course, even assuming her ideas about the Garden of Eden were true, her argument proved the exact opposite—that "the parts of the body" were so "vulgar and impure" they should only be discussed indirectly and allegorically.

In the United States, Woodhull needed a religious talk to justify what she was to say about human anatomy (particularly body excretions). In jaded and religiously cynical London and particularly before a crowd that wanted to be titillated, her pseudo-religious rationale wasn't necessary and could be left out. If she had given her quirky biblical interpretation in any detail, the critic writing for the *New York Times* would have brought it up for ridicule.

On the other hand, what she does say about the human body in the latter half of that American speech—remarks about digestion, circulation, elimination, and menstruation—were just the sort of things that would, as the correspondent suggested, disappoint those who attended her London speech hoping to hear about sex rather than body fluids.

It's also easy to understand why a writer for the *New York Times* might not mention any remarks she might have made about eugenics. At the *New York Times*, eugenics was not yet the fashionable cause it would become in the second decade of the next century. Even more important, judging by the published "The Garden of Eden," her remarks on eugenics may have been brief. She did not say much even in the lengthy printed version, and what she said was vague. Here is what she said on page 47 in the original, keeping these details in mind. The "savages" were Australian aborigines. "Prevention" referred to "means to avoid conception," later called birth control. "Spaying" meant the aboriginal practice of castrating men and removing the ovaries from women with disabilities such as deafness and blindness. "The remedy of the ancient Greeks" may refer to Spartan infanticide. If so, then Woodhull was hinting if the "unfit" do not listen to "educated public opinion," and have children, their babies might be taken away and killed. (In his 1905 *A Modern Utopia*, H. G. Wells, who took over Woodhull's role as eugenics popularizer, would claim that his future, "Utopia will kill all deformed and monstrous and evilly diseased births."[1]) Finally, while the second paragraph quoted below seems to be an introduction to "scientific propagation," it is not. In the next paragraph Woodhull turned back to bodily fluids and made a claim that menstruation excites the ovaries. Precisely as the *New York Times* reporter noted, this was a talk on the human body, but not the sort Woodhull's audience was expecting.

> The preceding indicates that certain savages have at least some regard for the future members of their community. It is to be hoped, however, that when the responsibilities of parentage become more fully recognized that neither method—preventives nor spaying, will be necessary; the force of educated public opinion will deter the unfit from propagating their kind: but if human beings have not sufficiently evolved to realize that the function of their generative organs is reproduction, I would recommend the remedy of the ancient Greeks.
>
> Scientific propagation is not a new idea which has originated with our nineteenth century civilization. (47)

The surgery had to be radical because surgical sterilization was still being developed. Woodhull's claim that "educated public opinion" might eventually remove the need for either method was a common blindness among some eugenists, as well as a ploy used by other eugenists to conceal their plans to make eugenics coercive. Even if normal people were educated to believe in eugenics, so they would never mate with someone who was 'unfit,' that did not mean the 'unfit' would not defy popular attitudes and mate among themselves for much the same reason that some defy public opinion and commit crimes. Eugenists tended to be so hostile toward those they thought 'unfit,' because they forgot such people could still think and feel. They weren't mere abstractions.

---

1. H. G. Wells, *A Modern Utopia* (New York: Charles Scribner's Sons, 1905) 129. In Michael W. Perry, *Pivot of Civilization in Historical Perspective* (Seattle: Inkling Books, 2001), 36.

In another passing reference to eugenics on page 53 in the original, Woodhull linked what she said about the Garden of Eden being the human body to the "law of evolution," combining her religious and scientific themes. Give man a "perfect physical body" through eugenics and "all other perfections follow," including the cure of social ills. That's nonsense. Handsome, healthy and intelligent people remain capable of evil, although they may be more difficult to catch and punish. But for the attractive and talented Woodhull, keeping imperfect, unfit, ugly people around and letting them reproduce, meant social perfection would never come. Note too her recognition that she is clashing with historic Christianity. Christianity focuses on saving the individual and eternal part of people, their souls, and thus offers hope for everyone, whatever the condition of their bodies. To perfect the body and create its "heaven," Woodhull's eugenics would have to reject most people as unworthy of the new humanity in the making. Woodhull even seems to think her remade humanity would never die.

That the law of evolution which makes growth the method by which intellectual altitude is reached, is also the law by which physical development goes forward; the perfected creation of man and his consequent salvation from death being physical and not moral, as has been falsely taught by almost the whole of Christendom. With a perfect physical body—man reconciled to God—all other perfections follow as its fruit, necessarily. The opposite proposition to this is the stumbling-block over which all Christians have fallen; they have given all their attention to saving the soul hereafter, when this salvation depends entirely upon saving the body here and now.

Woodhull would continue to lecture in London from time to time. For her later lecture series, she would place short advertisements in the *Times* (London), taking care to get the first entry in an often long listing. The second series would run for six nights in early 1878, January 11, 15, 18, and 22, as well as February 1 and 8. (All were on either a Tuesday or Friday evening). Here is an advertisement from Christmas Day of 1877 (page 6) for a "Second Series of Lectures." Additional advertisements on Monday, January 7, 1878 (page 8) and Monday, January 14 (page 12), would give her "startling ovation" the title "The Human Body the Temple of God." Her use of "series of lectures" suggests that not every night was the same, but doesn't make clear if she was giving six different lectures, two lectures three times, three lectures two times or what.

> VICTORIA CLAFLIN WOODHULL'S SECOND SERIES of LECTURES, St. James's (large) Hall, Jan. 11, 15, 18, 22, Feb. 1 and 8. Tickets for the series, one guinea. Address Prince's-gate Hotel, South Kensington.

Woodhull apparently felt that winter evenings attracted the largest audience. The following winter, she would give a "Third Series of Orations," advertising them in the December 19, 1878 issue of the *Times* (page 8). No specific dates were given for this 1879 series of "orations." She may have wanted to be free to adjust their number to how many people bought tickets.

> VICTORIA C. WOODHULL, from America, will DELIVER her THIRD SERIES of ORATIONS, The Human Body the Temple of God, during February and March. All communications to be addressed St. James's-hall, Piccadilly.

Since I was unable to find any news articles about Woodhull in the *Times*, it seems likely that at this time she was being snubbed as being unworthy of comment. That sort of disdain may have bothered her more than the ruder comments of her more ordinary critics.

Finally, fourteen years later, on Friday, March 24, 1893, Woodhull advertised a single Friday night "Address" on a theme that was clearly eugenic, although the second title is the same as for her two previous series. She also seemed be shifting her means for attracting an audience from notoriety to respectability. Notice that, having lived in England for some sixteen years, she no longer describes herself as "from America" and includes her eminent married name underneath her literary name.[2]

> ST. JAMES'S-HALL.—VICTORIA WOODHULL (Mrs. John Biddulph Martin) Will DELIVER an ADDRESS on The Scientific Propagation of the Race, or The Human Body the Temple of God, THIS (Friday) EVENING, at 8 o'clock. Tickets of the usual agents, and at Tree's office, St. James's-hall.

The first title, "The Scientific Propagation of the Race," is almost identical to that of Chapter 8 in this book, so it's likely that the content was similar. Some may suspect this 1893 speech is the one Woodhull meant when she spoke the 1912 *Evening Standard* about an 1877 speech on "The Scientific Propagation of the Human Race." That's unlikely, since Woodhull also refers to staying as the "guest of Lord and Lady Mount Temple" and Lord Temple died on 16 October 1888. The fact that he did not become titled (as Baron) until 25 May 1880, may set early limit on the date of that speech, but that's not strictly required. (The Lord and Lady were important social figures, so there may be historical records of their assistance with Woodhull's speech.)

The rest of this chapter is an exact facsimile of Woodhull's "The Garden of Eden" as it was published as the first chapter of her 1890 book, *The Temple of God*. As she noted in the book's preface (quoted in Chapter 3), the book contained many of her speeches from the 1870s. This is one of them.

---

2. So conscious of her respectability, was she, that the following year she sued the British Library for libel merely because it had in its collection a rarely checked out book about her 1872 'free love' controversy with the Rev. Henry Ward Beecher, a prominent liberal Congregationalist pastor. See the January 24, 1894 issue of the (London) *Times*, page 3.

THE GARDEN OF EDEN;

or,

THE PARADISE LOST AND FOUND.

INTRODUCTION.

Most of the ideas which permeate our social, religious, and political institutions of to-day arise from misconceptions of the human body. These institutions which are the outcome of civilization define laws to regulate and control the actions of *human beings*; and yet, the proper understanding of the growth and development of man individually was, and is, considered of secondary importance in adjusting these laws. My philosophy has been on the lines of Aristotle, who said, " The *nature* of everything is best seen in its smallest portions." My efforts were for the individual or ontogenic development of humanity as the only basis upon which to frame any laws—that by understanding and giving the proper attention to this the *quality* of the whole must of necessity ultimately reach a higher standard. And as the influence of woman is vital, no advance could be made until the co-operation of woman was properly understood and insisted upon as essential to any ideal society, to any true realization of religion, to any perfect government. Active not passive aid is what I demanded from woman. She must be appreciated as the architect of the human race. Men are what their mothers make them. Their intelligence or ignorance has the power to teach them to revere or desecrate

4

womanhood. Night after night throughout the United States I pleaded for the intellectual emancipation and the redemption of womanhood from sexual slavery—insisting that social evils could only be eliminated by making your daughters the peers of your sons—that the greatness of a nation depends upon its mothers. I denounced as criminal the ignorant marriages which were filling the world with their hereditary consequences of woe, shame, and every manner of crime. The theme of my public work was that I would make it a criminal offence to allow persons to marry in ignorance of parental responsibility. I realized that the Bible was little understood, but had in it the germ of a great and divine truth—that is the redemption of the body. A part of this truth regarding the "Garden of Eden," &c., I gave in my extemporaneous lectures. It was afterwards put into consecutive biblical articles and pamphlets. I did not then give the whole truth with which my soul had become illuminated; for I knew the fulness of time was not yet. I considered the work I was then doing as a necessary part of the evolution of thought—as initiatory to my reformatory work. In a book that I am at present writing, it is my intention to give the entire truth of all Bibles, which was only partially understood by primeval religious sects through their ignorance of the phenomena of life.

V. C. W. M.,
17, Hyde Park Gate, London.

"BUT IN THE DAYS OF THE VOICE OF THE SEVENTH ANGEL, WHEN HE SHALL BEGIN TO SOUND, THE MYSTERY OF GOD SHALL BE FINISHED."
Revelation x. 7.

The First Book of Moses, called Genesis.

Chapter II.

Thus the heavens and the earth were finished, and all the host of them.

2 And on the seventh day God ended his work which he had made; and he rested on the seventh day from all his work which he had made.

3 And God blessed the seventh day, and sanctified it: because that in it he had rested from all his work which God created and made.

4 ¶ These *are* the generations of the heavens and of the earth when they were created, in the day that the LORD God made the earth and the heavens,

5 And every plant of the field before it was in the earth and every herb of the field before it grew: for the LORD God had not caused it to rain upon the earth, and *there was* not a man to till the ground.

6 But there went up a mist from the earth, and watered the whole face of the ground.

7 And the LORD God formed man *of* the dust of the ground, and breathed into his nostrils the breath of life; and man became a living soul.

8 ¶ And the LORD God planted a garden eastward in Eden; and there he put the man whom he had formed.

9 And out of the ground made the LORD God to grow every tree that is pleasant to the sight, and good for food: the tree of life also in the midst of the garden, and the tree of knowledge of good and evil.

10 And a river went out of Eden to water the garden; and from thence it was parted, and became into four heads.

11 The name of the first *is* Pison: that *is* it which compasseth the whole land of Havilah, where *there is* gold;

12 And the gold of that land *is* good: there *is* bdellium and the onyx stone.

13 And the name of the second river *is* Gihon: the same *is* it that compasseth the whole land of Ethiopia.

14 And the name of the third river *is* Hiddekel: that *is* it which goeth toward the east of Assyria. And the fourth river *is* Euphrates.

15 And the LORD God took the man, and put him into the garden of Eden to dress it and to keep it.

16 And the LORD God commanded the man, saying, Of every tree of the garden thou mayest freely eat:

17 But of the tree of the knowledge of good and evil, thou shalt not eat of it: for in the day that thou eatest thereof thou shalt surely die.

18 ¶ And the LORD God said, *It is* not good that the man should be alone; I will make him an help meet for him.

19 And out of the ground the LORD God formed every beast of the field, and every fowl of the air; and brought *them* unto Adam to see what he would call them; and whatsoever Adam called every living creature, that *was* the name thereof.

20 And Adam gave names to all cattle, and to the fowl of the air, and to every beast of the field; but for Adam there was not found an help meet for him.

21 And the LORD God caused a deep sleep to fall upon Adam, and he slept: and he took one of his ribs, and closed up the flesh instead thereof;

22 And the rib, which the LORD God had taken from man, made he a woman, and brought her unto the man.

23 And Adam said, This *is* now bone of my bones, and flesh of my flesh: she shall be called Woman, because she was taken out of Man.

24 Therefore shall a man leave his father and his mother, and shall cleave unto his wife: and they shall be one flesh.

25 And they were both naked, the man and his wife, and were not ashamed.

Chapter III.

Now the serpent was more subtle than any beast of the field which the Lord God had made. And he said unto the woman, Yea, hath God said, Ye shall not eat of every tree of the garden?

2 And the woman said unto the serpent: We may eat of the fruit of the trees of the garden:

3 But of the fruit of the tree which *is* in the midst of the garden, God hath said, Ye shall not eat of it, neither shall ye touch it, lest ye die:

4 And the serpent said unto the woman, Ye shall not surely die:

5 For God doth know that in the day ye eat thereof, then your eyes shall be opened, and ye shall be as gods, knowing good and evil.

6 And when the woman saw that the tree *was* good for food, and that it *was* pleasant to the eyes, and a tree to be desired to make *one* wise, she took of the fruit thereof, and did eat; and gave also unto her husband with her, and he did eat.

7 And the eyes of them both were opened, and they knew that they *were* naked: and they sewed fig leaves together, and made themselves aprons.

8 And they heard the voice of the Lord God walking in the garden in the cool of the day: and Adam and his wife hid themselves from the presence of the Lord God amongst the trees of the garden.

9 And the Lord God called unto Adam, and said unto him Where *art* thou?

10 And he said, I heard thy voice in the garden, and I was afraid, because I *was* naked: and I hid myself.

11 And he said, Who told thee that thou *wast* naked? Hast thou eaten of the tree, whereof I commanded thee that thou shouldest not eat?

12 And the man said, The woman whom thou gavest *to be* with me, she gave me of the tree, and I did eat.

13 And the Lord God said unto the woman, What *is* this *that* thou hast done? And the woman said, The serpent beguiled me, and I did eat.

14 And the Lord God said unto the serpent, Because thou has done this, thou *art* cursed above all cattle, and above every

8

beast of the field: upon thy belly shalt thou go, and dust shalt thou eat all the days of thy life:

15 And I will put enmity between thee and the woman, and between thy seed and her seed; it shall bruise thy head, and thou shalt bruise his heel.

16 Unto the woman he said, I will greatly multiply thy sorrow and thy conception; in sorrow thou shalt bring forth children; and thy desire *shall be* to thy husband, and he shall rule over thee.

17 And unto Adam he said, Because thou hast hearkened unto the voice of thy wife, and hast eaten of the tree, of which I commanded thee, saying, Thou shalt not eat of it: cursed *is* the ground for thy sake; in sorrow shalt thou eat *of* it all the days of thy life;

18 Thorns also and thistles shall it bring forth to thee: and thou shalt eat the herb of the field:

19 In the sweat of thy face shalt thou eat bread, till thou return unto the ground; for out of it wast thou taken: for dust thou *art,* and unto dust shalt thou return.

20 And Adam called his wife's name Eve: because she was the mother of all living.

21 Unto Adam also, and to his wife, did the Lord God make coats of skins, and clothed them.

22 ¶ And the Lord God said, Behold, the man is become as one of us, to know good and evil: and now, lest he put forth his hand, and take also of the tree of life, and eat, and live for ever;

23 Therefore the Lord God sent him forth from the garden of Eden, to till the ground from whence he was taken.

24 So he drove out the man; and he placed at the east of the garden of Eden cherubims, and a flaming sword which turned every way, to keep the way of the tree of life.

I take up this book and call your attention to it. You perhaps will say, "Oh, that is the old Bible, worn threadbare long ago. We do not wish to be fed with its dry husks. We want living food and drink." Well, that is what I am going to give you.

Yes! it is an old book, a very old book. There are very few books extant that can compare with it, on the score of age, at least. Some parts of it were written over three thousand years ago; and all of it more than eighteen hundred years ago. Yes! an old book. And yet everybody seems to have one about the house. What is the matter with the old book? Why do people cling to it with such tenacity? Can any of those who have laid it on the shelf as worthless answer these questions? Why do they not burn it, so that it shall no longer cumber the house? This was a mystery to me for many years; but it is so no longer. I know the reason for its hold upon the people. It contains that, though clad in mystery, which acts upon the soul like a potent spell; like a magnet, which it is indeed. Had it no value, or had its value been wholly extracted; were there no truth in it unrevealed, it had long since ceased to exert any influence whatever over anybody. Books that are exhausted of their truth by its being transferred to the minds of the people, lose their force and die. And this is the reason that I ask you to search its hidden mystery with me; to cast aside preconceived ideas of its meaning; to commence to read it as if it were for the first time.

Religion and science admit that there was an original cause which set up in matter the motion that ultimated in man. The latter examines into the various works that preceded his appearance, and discovers that he came as a result of them all; indeed, that, except they had first existed, he could never have lived; that the omission of a single progressive step in the creative plan would have defeated the work. But science goes further than this. It not only asserts that man was the last link in a long chain of development, but it also maintains that, when the creation once began, there was no power residing anywhere that could have interposed its edicts to stay the progress, or defeat the final production of man; that he was a necessary product of creation, as fruit is of the tree; and that all the designs and purposes of the moving power were contained in and exhausted by his creation; that is, that as a fruit of the creative plan, man was the highest possibility of the universe.

Religious theory, in inquiring into the creation of man, has pursued the method precisely the reverse of this. Having found man on the earth, it assumes that he was a special creation: that is, that God, having purposed in Himself that He would create man, set Himself about to prepare a place in which he was to live; the earth, formed according to the account in Genesis, being that place. I say that this is the theory of religionists: but it is by no means certain that their account of the creation justifies any such conclusion. The biblical account of the creation is an allegorical picture of it, which, in detail, is strikingly in harmony with the real truth. "In the beginning God created the heavens and the earth, and the earth was without form, and void." There were light and darkness—day and night. There were the divisions into water and land; the vegetation, fish, fowl, beast, and man; and next, the rest from labour. In so few words, who could make a clearer statement of what we know about the creation of the earth than this?

We must remember that the Bible does not pretend to be a scientific book at all. It deals altogether with the inspirational or spirit side of the universe. St. Paul informs us that the God of the Bible "is a spirit." At least the translators have made him state it thus; but it is not exactly as he wrote it, although in the end it has the same significance, since if God is a spirit, a spirit is also God. The original Greek of this, which is what Paul meant to say, and did say, and which is the truth, religiously and scientifically also, is *Pneuma Theos* Pneuma meaning spirit, and Theos God. According to St. Paul, then, spirit is God, and according to science, the life that is in the world is its creative cause; so both agree in their fundamental propositions, however much the priestcraft of the world may have attempted to twist St. Paul into accordance with their

11

ideas of the personal character of God, and in placing God first in the declaration, instead of making spirit the predominant idea. The biblical Creator, then, as defined by the Apostle, is spirit: "And the Spirit of God moved upon the face of the waters" (Genesis i. 2), which was the beginning of creation. The fact, stated scientifically, would be: And the power (or the spirit) resident in matter, caused it to move, and by this motion the earth began to assume form and to be an independent existence, revolving upon its own axis as a planet, and around the sun as its centre.

But I do not purpose to enter into a detailed discussion of the relations which the Bible creation bears to the demonstrations of geology and astronomy. I desire to show merely that the Bible Creator, God, is not at all incompatible with the power which science is compelled to admit as having been the creative cause of all things.

If we take the Darwinian theory and endeavour to find where and how man came, we are led necessarily to a time when there was nothing existing higher than that type of animal by which man is connected with the brute creation, and through which he came to be man. Man is an animal; but he is something more as well. He knows good and evil, and this is to be more than an animal. There was a time, however, when man did not know good and evil. It was then that the form—the human man—was in existence; and it is easy to conceive that the whole face of the earth may have been occupied by human beings who were nothing more than animals, as it is now occupied by them being more than animals. These were the male and female whom God created according to the first chapter of Genesis. It does not mean at all that they were a single male and female. They were not Adam and Eve then. They were simply male and female man, or Adam; for in chapter v. verse 2, we are told, "Male and female created he them and called *their* name Adam;" that is, the human animals that inhabited the earth were called Adam.

Now, this is precisely the condition in which science informs us that man, at one time, must have been. He was not created at one and the same time, physically, mentally, and morally; he may have lived for ages in this animal condition. Of this, Moses tells us nothing in his history of the creation. But as there were immense periods of time—days—between the various epochs of the creation of which he tells us nothing, we must remember that with God there are no divisions of time, for all time is eternity. But there came a point in time when male and female man had developed to the condition in which the gleams of reason began to light up the horizon of the intellect, as the first rays of the morning sun lights the tallest mountains which reflect them into the valleys below.

It was at this time that the Lord God " planted a garden eastward in Eden," in which he put the man whom he had formed " to dress it and to keep it." It is sufficient here to say that it consisted of the ground that was cursed by reason of the sin that Adam and Eve committed. Nor is it essential to the argument, at this time, to consider whether this ground—this garden—was a single one, or whether there was more than one, scattered here and there among male and female men.

The probability is, however, that these names refer to *conditions* and not to individuals. Indeed, it may as well be said now, as later, that the Bible is not a history of individuals and nations at all, but rather the condition and development of universal man, sometimes, perhaps often, using historical facts by which to typify them, but for all that, intended to refer to the *interior* instead of the exterior progress of man; that is, the Bible relates to the building and progress of God's holy temple.

It is upon the consequences of the fall of man, which is therein set forth, that the necessity for a plan of redemption rests. Take away the first three chapters of Genesis and the superstructure of orthodox religion would topple and fall. So, then, it becomes necessary, since Christians have made them vital, to inquire into what these chapters mean—to inquire what was the

Garden of Eden, there so graphically set forth—whether a spot of ground situated somewhere on the surface of the earth, or something altogether different—something, perhaps, that it may seldom or never have been suspected of being, and yet something that the language of these chapters plainly states it to have been; or, as may prove to be the exact truth, something other than which it is impossible to derive from the language in which the description is clothed. *For instance, if the various parts of a thing be described as parts, when the parts are put together, that which they form must be the real thing which was in the mind of its relator.* Therefore, if when we shall take the several things described by Moses and put them together, they shall be found to constitute something widely different from a spot of ground on the surface of the earth, why then we shall be forced to conclude that it was not such a spot that Moses had in view when he wrote the second chapter of Genesis; and therefore, also, that the Garden of Eden must be sought elsewhere than in a geographical location.

Indeed, I do not hesitate to say here at the outset, knowing full well the responsibility of the assertion, that I can demonstrate to you—to any minister or number of ministers—to all the theologians everywhere—that there is not a shadow of reason contained in the language used for concluding that the Garden of Eden ever was a geographical locality; but, on the contrary, without resorting to anything outside of the Bible—without any words of my own—I can show, beyond the possibility of cavil, and to the satisfaction of all who will give me their attention, that the Garden of Eden is something altogether different from a vegetable patch, or a fruit or flower garden; aye, more definite than this still—that I can demonstrate, so that there can be no manner of question about it, just what this garden was, and what it still is, with its cherubim and flaming sword defending the approach to its sacred precincts. Nor, as I said, will I go outside of the Bible to do all this, so that, when it shall be done, none can say that I have cited any irrelevant matter or any questionable authority.

The Bible has seldom, if ever—certainly never by professing Christians—been searched with the view to discover any new truth that might not be in harmony with their preconceived ideas as to what the truth ought to be; that is to say, it has never been searched fearlessly of what the truth might prove to be. The seal of mystery that is visible all over the face of the Bible, and that is clearly set forth in words within itself, has never been broken, nor the veil penetrated which hides its real significance from the minds of the people; while the attempts that have been made to interpret this significance have had their origin in a desire to verify some already entertained idea.

To want the truth for the sake of the truth—to want the truth, let it be what it may and lead where it may has had, so far, no conspicuous following in the world, or at least so few that, practically, it may be said that there has never been any desire for the truth for its own sake. When the truth has appeared to be in antagonism with the cherished conceits of the people, they have shut their eyes and closed their hearts against it, and blocked up all avenues for its approach to them. One of the best evidences that the full truth is soon to dawn upon the world, lies in the fact that there are now a few persons who want the truth for its own sake, and who will follow it wherever it may lead them.

For one I want the truth, the whole truth; and I will proclaim it, no matter if it be opposed to every vestige of organization extant—political, social, religious! No matter if it be revolutionary to every time-honoured institution in existence! Let creeds fall if they will; let churches topple if they must; let anarchy even reign temporarily if it cannot be avoided, but let us for once in the world have the simple, plain truth; and let us welcome it because it is the truth, and not because it may or may not be in accord with popular notions and opinions.

But now to the Garden of Eden: In the second chapter of Genesis, beginning at the 8th verse, and, for the present, ending with the 14th verse, we read thus:—

8 "And the Lord God planted a garden eastward in Eden; and there he put the man whom he had formed.

10 "And a river went out of Eden to water the garden; and from thence it was parted, and became into four heads.

11 "And the name of the first river *is* Pison: that *is* it which compasseth the whole land of Havilah, where *there is* gold; there *is* bdellium and the onyx stone.

12 "And the gold of that land is good.

13 "And the name of the second river *is* Gihon: the same *is* it that compasseth the whole land of Ethiopia.

14 "And the name of the third river *is* Hiddekel; that *is* it which goeth toward the east of Assyria. And the fourth river *is* Euphrates."

These six verses comprise the physical description of the garden, and it is upon them that the structure, now to be taken in pieces and examined, rests. For a moment let us look at the language in its literal sense and see whether in this way it appears as if it were probable even that it may be true. "And the Lord God planted a garden eastward in Eden." Bible geographers and commentators say that the locality of the garden is lost, and they do not pretend to tell where Eden is, or was, to say nothing about a particular spot in Eden where the garden was planted. It is supposed that Eden was somewhere in Asia; in fact, somewhere in the neighbourhood of Jerusalem, the holy city. If they who say so knew how nearly they have hit upon the truth without knowing what the truth is, the ministers would indeed be astonished. But where is eastward in Eden? Since the best informed Christian geographers can give no help to aid us in the search which we propose to make for this famous garden, we might as well conclude that it is anywhere else in the world as to conclude that it is in Asia.

But an astute person suggests that it must have been in Western Asia, because the rivers named as being in the garden are there. Yes! There were some rivers, and there were some countries in which they were situated, and yet we are coolly informed that the garden is lost, as if it were a matter of only the slightest moment. But will Christians assert, with the expectation that it will be believed, that the location of the four rivers and of the countries in which they were located, are lost with the garden. To say that the garden is lost is virtually to say just that. The four rivers are enumerated specifically, to wit: the Pison, the Gihon, the Hiddekel, and the Euphrates. Are these rivers lost and also the countries Havilah, Ethiopia, and Assyria—all well known geographical terms? If they are not, how does it happen that the garden can be lost? There seems to be something very strange about all this.

And as the allegory continues, when the Lord God had expelled Adam from the garden, we are informed that he "placed at the east of the garden Cherubims [*the Cherubims, the eyes; and the flaming sword, the tongue*], and a flaming sword which turned every way to keep the way of the tree of life." Is it not proper also to inquire after these sentries of the Lord God? What has become of them, and the tree of life that they were set to guard? If they were set "at the east of the garden," and the garden was in Western Asia, why are they not to be found somewhere now? If I were anxious about the consistency of my theology, I should send off a Livingstone at once to hunt up this garden, fearing lest my religion might go to keep company with the garden upon which it is founded. I will venture the opinion that anyone who should start upon that journey, would have a more difficult task than discovering the sources of the Nile, or the North Pole, has proved to be.

But what about that tree of life which was in the midst of the garden? What has become of that? Is that lost also? Is that perished? and if so, are there any more in the world? The Lord God expelled Adam from the garden "lest he should put forth his hand and take of the tree of life, and eat and live for ever." It seems that this kind of tree was not very common then, at least in that part of the world. If they are common in any part, I have never heard of them. If there were any in existence, it is my opinion that two cherubims and one flaming sword would afford them but poor protection against the ravages of a people who cling to life with the tenacity with which most of the people exhibit, not excepting that portion which believes itself safe from the uncomfortable regions of the other world, and who should most desire to die.

Thousands of the wisest men of Oriental nations have searched Asia over and over, and have failed to find a single tree of life anywhere. Has the logic of this fact ever had its legitimate weight in the consideration of this matter? I think not. The generality of people have never thought upon this subject at all, or about anything else connected with their religion.

In the second chapter of Genesis we are told all about the countries in which the garden was located, and the rivers that bounded it. From what I have already said, however, it is understood that I do not believe in this garden as commonly understood; nor do I believe that so important a spot as this garden is claimed to be, should be summarily given up as lost. The most important clue is the course of one of the rivers of this garden. Let us follow it to its source; for, in the tenth verse, it says, "And a river went out of Eden to water the garden; and from thence it was parted, and became into four heads; that is to say, it gave off four branches. Let us see which of the four rivers we shall select as the basis of operations, and on which to make the ascent to find the place where it divides from the main river. The first river, as we have seen, is called Pison. As we can find no geographical mention of this river, we shall be obliged to omit Pison. The next in order is the Gihon. We are told (2 Chronicles xxxii. 30) that King Hezekiah turned the upper water-course of Gihon so that it should run by the City of David. That ought to be definite; but we fear, if we were to go to the City of David to-day, we should find the river in the same condition as the garden itself which it once watered—that its location is lost. So we must also pass the Gihon, and turn to the next, which is Hiddekel. Though both Moses and Daniel said that this river was in Assyria, we can find no geographical mention made of its locality anywhere; therefore we shall be obliged to dismiss this with the others, and have recourse to the last one, which is the Euphrates. We all know where the Euphrates river is located, and if we can reach its banks, and follow up its course, we must, as a matter of necessity, find its source; and in finding it, find also the greater river Pison, from which it divides. Having done this, all the other rivers also will be discovered. There can be no mistaking the place, since it was at that point where the great river divided into four heads. When we arrive at this place, we shall be, at least, near the garden.

But, alas for our hopes! We wander along the banks of the beautiful Euphrates, from its mouth to its source, and find no place where it divides from another river; but, on the contrary, discover a number flowing into its ever-increasing stream. And now we cross to the opposite shore, and again from the Persian Gulf to the mountains of Armenia, seek the desired spot, but still are doomed to disappointment. If this be the river Moses describes, then his description is not true. The Euphrates river does not divide from any other river, but has its own source, as other rivers have their sources. So our last hope from the rivers is gone. We must dismiss the Euphrates as well as the Pison, the Gihon, and the Hiddekel.

Let us not, however, be altogether discouraged by our repeated failures with the rivers. The object in view is too important to be hastily abandoned. We have not yet exhausted our means of discovery. So, with heavy hearts, we will turn our backs upon the rivers, and seek elsewhere, hoping for better success. Since we cannot find the garden through the medium of its rivers, perhaps, if we reverse the process, we may be able to hunt up the rivers by seeking for them in the countries in which Moses said they were located. The river Pison, so he informs us, is the name of the first of the four rivers into which the great river divides, and "that it compasseth the whole land of Havilah." Now, certainly, we ought to be able to find the river Pison, for Havilah is a district of country on the Red Sea, in Arabia, south-east of Sanaa. Referring to the map of this portion of the earth, we readily find the land called Havilah. But what is this? It is not an island at all. Moses said that it was compassed—that is, encircled—by the river Pison, and that should make it an island. But there is no river that runs about this Havilah. Indeed, there is not any river in this land that is laid down on the maps. Moreover, we find from the conformation of this land that it is a physical impossibility for a stream of water to compass it. The western part of Havilah rests upon the Red Sea, where no river could ever have run. So it cannot be said that there might have been a river there in the time of Moses, which has since disappeared. He must have been very much mistaken, or else the land of Havilah, to which he referred, is something quite apart from geographical land; and yet Moses is most explicit, since he says that the ground of this land was cursed.

Having failed with Havilah, we will go on to the next. "And the name of the second river is Gihon," says Moses, " the same is it that compasseth the whole land of Ethiopia." Now, Ethiopia is a large country—a very large country—and Moses says that the whole land was compassed by this river Gihon; a river that should encircle the whole of this land of Ethiopia must be no less than three thousand miles in length. It were impossible to lose such a river as this; hence, if it ever had an existence anywhere, it must be now in existence somewhere. Besides it must have been a still larger river even than this from which so large a one could have been given off. But what is this that we find? Ethiopia is a vast domain, situated in the very heart of Africa, with mountains on the north, mountains on the east, mountains everywhere. If the second chapter of Genesis is geography, Moses must have meant to have said there were mountains instead of a river compassing the whole land of Ethiopia, or else his Ethiopia was some country other than the one which we have under consideration, and one of which there is nothing known in our day save what Moses tells us.

We will now re-cross the Red Sea into Asia, and go through the land of Assyria, looking for the river Hiddekel, which Moses says is there. Turning again to the maps, we also again fail to find such a river as Hiddekel there set down, and we run through the geographies fruitlessly. As far as our investigations have been pushed, we can find two places only in all the books where this river is mentioned, and these occur in the text, and in Daniel x. 4. This is the river on the banks of which Daniel had the most remarkable vision recorded in the Old Testament; and it rises into the greatest significance by reason of the character of that vision. Where should this river be? Bible geographers endeavour to account for the discrepancies between the Bible and the geographies by saying that it is supposed that this river Hiddekel was the one now known as the Tigris. To be sure the Tigris runs with a swift current as did the Hiddekel; but it is not in the right place, nor does it run in the right direction. The maps show that the river Tigris instead of running "to the east of Assyria," runs southward into the Persian Gulf. Nor do the maps discover any river running to the east of Assyria which may be taken for the river Hiddekel of the Bible. So we shall have to abandon the search for the Garden of Eden. We have exhausted the rivers, and the countries also, in which Moses set it down as being located.

Although we have not discovered the garden, we have found all the countries named by Moses. If the Garden of Eden really consisted of all of these countries, and for some reason, now unknown, their rivers cannot be discovered, it must have been a very large garden—almost as large as the half of North America. But we have stumbled upon one rather singular fact that needs to be explained: We know that the river Euphrates is in Turkey in Asia. Then how does it happen that another river, which has its source in the same river from which it is said to divide, is in Ethiopia, in Africa—which is separated from both Assyria and Havilah by the Red Sea? How does the river Gihon find its way across the Red Sea into Ethiopia to compass the whole of that land? Failing to explain this, however, an attempt perhaps will be made to clear it away upon the well-known hypothesis, that with God all things are possible; and consequently, that it was possible for Him to construct a river that could run under the Red Sea to get into Ethiopia; and a garden made up of large countries, widely separated each from the other, and still be altogether in one place, with a single tree in its midst; to watch and guard which, cherubims and a flaming sword were set at the east of the garden, a distance of not less than three thousand miles from its western limits.

But why dwell longer upon this mass, geographically considered, of physical impossibilities and absurdities. Any school boy of twelve years of age who should read the description of this garden and not discover that it has no geographical significance whatever, ought to be reprimanded for his stupidity. Nevertheless, learned Divines have written and preached for ages over this mythical garden just as if it ever had a geographical existence, and never suspected that what they were writing and talking about was all a fable, simply incredible.

Geography must have been interdicted in the schools where they were educated; or else the theological spectacles must have been so highly coloured by authority that they could not perceive that the geography of the Bible and that of the face of the earth ought to agree somewhat, which in this case it does not at all.

Do you not begin to see how preposterous and impossible, how

22

contradictory and absurd, it is even to pretend to think that the Garden of Eden is a geographical locality? I challenge any clergyman—all clergymen—to impeach the truth, force, or application which I shall make of a single one of the rivers and countries of this famous garden. And I call upon them, failing to do it, to lay this whole fable open to their people as I have laid it open to you. Will they do so? If they care more for their theology than they do for the truth, No! But if they love the truth better than they do their theology, Yes!

But was there not a Garden of Eden! I think some will query in their minds. Or is this thing a bare-faced fraud upon the credulity of a simple people? Oh, yes!—There was a Garden of Eden. It is not at all a fraud. The fraud has been in the preachers, who would not look into the Bible with sufficient reason to discover a most palpable absurdity. There is where the fraud lies, and there it will, sooner or later, come to rest. I do not say that they have done this intentionally. I say only that they have done it; and the responsibility for having misled the people, year after year for centuries, rests with them. They have been the blind leading the blind; and they have both fallen into the ditch of deception.

It was necessary, before there could be a successful search to find the Garden of Eden, to clear away the last vestige of possibility upon which to conceive that it might have been a geographical locality. Have I not made it clear to you all that it was not? If I have, then we are ready to look without bias or prejudice in other directions to find it—for there was a Garden of Eden.

As introductory to this part of my subject, it is proper to say that the general misunderstanding of the real meaning of the Bible can be easily explained. The proper names have been translated from the original languages, arbitrarily, and mingled with the common usage of the new languages, in such a way as to deprive them of their original significance, unless we are familiar with the meaning of the words from which they were translated. The term Eden is a good example. If we are ignorant of the meaning of Eden, in the original language, its use signifies to us that there was a garden which bore this name simply for a designation. But if we were to use the meaning of the word, in the place of the word itself, then we should get at the meaning of the one who gave this designation to the garden. The failure to translate the Bible after this rule is one reason for its still being veiled in mystery: and this fact will become still more evident when it is remembered that, in early times, names were given to persons and things, not merely that they might have a name, but to embody their chief characteristics.

So, then, the first step to be taken is to inquire into the significance of the names that the rivers and countries of the Garden of Eden bear. I cannot explain better what I mean by this than by quoting St. Paul on this very subject. In his letter to the Galatians, beginning at the 22nd verse of the 4th chapter, he says:—

"For it is written, that Abraham had two sons; the one by a bondwoman, the other by a freewoman. But he who was of the bondwoman was born after the flesh; but he who was of the freewoman was by promise. Which things are an allegory: for these are the two covenants; the one from the Mount Sinai, which is Agar. For this Agar is Mount Sinai in Arabia, and answereth to Jerusalem, which is now in bondage with her children."

Now, suppose that Paul had not entered into any explanation about this story regarding Abraham. Of course we should have been left to suppose, conjecturing after the manner of the suppositions about the Garden of Eden, that Abraham really had these two children as described; and so he did. But Paul says it is an allegory; meaning that they represented all children born under both covenants: those of the first being children of bondage—that is, born in sin—and those of the latter being free-born, or born free from sin. This is still more evident when the last verse quoted is interpreted. Jerusalem always means woman, and to get the meaning of the verse it should be read thus: For this Agar is Mount Sinai in Arabia, and answereth to "*woman*," who is in bondage with her children. The succeeding verse demonstrates this clearly, since it reads: "But Jerusalem [woman, remember], which is above, is free, which is the mother of us all." The interpretation of the meaning of the words used in the description of the Garden of Eden will make equally as wonderful transformations of the apparent meaning as are made by Paul in this allegorical story about Abraham.

It is now generally admitted that the account of the creation contained in the first chapter of Genesis is wholly allegorical. Having admitted so much, it would be preposterous to not also conclude that the allegory extends into the second chapter, and includes the Garden of Eden. If the first chapter refers to the creation of the physical universe, it is not too much to say that it is a wonderfully correct picture of the manner in which the world was evolved. If we apply the same statement to the second chapter, then we are ready to inquire what the subject is which this allegorical picture represents.

First in the allegory is the name of the garden, then its rivers, and lastly the countries through which they run. Passing, for the time, the name of the garden, we will begin by inquiring into the rivers. The name of the first is Pison; that of the second is Gihon; and that of the third is Hiddekel; and that of the fourth is Euphrates. These were the names of all the rivers mentioned as being in the garden. Turning to Cruden's Concordance, quarto edition, there will be found what is called "An alphabetical table of the proper names in the Old and New Testaments, together with their meaning or signification in the original languages." That is what we want. And the study of it will convince everybody of what and where the Garden of Eden is, and make it clear why its locality has been lost, as superficial students of the Bible say it has.

In that learned work we read thus: "Pison—changing or doubling, or extension of the mouth."

"Gihon—The Valley of Grace, or breast, or impetuous." In other authorities this word is held to mean "Bursting forth as from a fountain, or from the womb."

"Hiddekel—a sharp voice or sound;" other authorities say, "Swift, which refers to the swiftness of the current."

"Euphrates—that makes fruitful or grows." Now we may inquire into the meaning of the names of the countries in which these rivers were situated.

"Havilah—that suffers pain, that brings forth."

"Ethiopia—Blackness—[Darkness]—heat, burning."

Assyria is the country of, and signifies Ashur, "One that is happy," which would make the meaning of Assyria to be, the land of the happy; or the land in which the happy dwell.

And the whole of these rivers and countries combined form the Garden of Eden, which, as we learn, means: "Pleasure or delight." So, the Garden of Eden into which the Lord God put the man whom he had formed, "to dress it and to keep it," was the garden situated in the land of pleasure or delight. Remember that these words are not mine, but that I quote them from that acknowledged authority, Cruden's Concordance.

It will be necessary to give the meaning of one more word before entering upon the application of the meaning of these words, and that is "East." The direction of east is always to the light, let the light be of whatsoever kind—physical, mental, or moral. Toward the west means going, following, or looking after the receding light. These are astrologic terms, and were taken from the ancient magi, who derived them from the sun. When the light of the sun is looked for as coming, it is toward the east that the eye is turned, because it always comes from that direction; but when we look toward the west to observe it, it is to see the departing light which precedes darkness. So, east, in our investigations, means toward or into the light. We look, allegorically, to the east when we seek a new light or a new truth. The Star in the East, which stood over the place where "the young child lay," was the new spiritual light that came by Him into the world. The same meaning attaches to the word east wherever it appears in the Bible.

"And the Lord God planted a garden eastward in Eden, and there he put the man whom he had formed." The signification of these words would make the text read thus: "And the Lord God planted a garden in pleasure or delight, the fruit of which was to be, or was, a new revelation in, or a new light to, the world."

"And a river went out of Eden to water the garden and from thence it was parted and became into four heads. The name of the first river was Pison; that is, it which compasseth the whole land of Havilah, where there is gold." If this language be transposed into the signification of its words it would read thus: "And a river went out of the garden in which there is pleasure or delight, which river watered, fed, and drained the garden; and to water, feed, and drain the garden it was divided into four channels. The first of these new rivers, and the main one in which all the others found their sources, was the extension of the mouth; and as this river ran onward in its course, compassing or encircling that which suffers pain and brings forth fruit, the character of its waters was constantly changing by reason of its giving food and receiving refuse from the land through which it ran; and in this land there were things of great value, besides the bdellium and the onyx-stone."

This is the full meaning of the 10th, 11th, and 12th verses of the second chapter. The 13th verse reads thus: "And the name of the second river is Gihon; the same is it that compasseth the whole land of Ethiopia." This, transposed into its signification, would read thus: "And the second river of the garden bursts forth as a fountain, or from the womb, from the valley of grace, in which valley it flows in darkness and in heat."

The first clause of the 14th verse reads thus: "And the name of the third river is Hiddekel; that is it which goeth toward the east of Assyria." The translation of this, into its signification, would be as follows: "The third river of the garden runs with a swift current and a sharp sound into the light. Furthermore, this river, being in that part of the land known as Mesopotamia, which, interpreted, means 'in the midst of the rivers,' is surrounded by the other rivers of the garden, and is, therefore, situated in their midst."

The last sentence of the 14th verse is: "And the fourth river is Euphrates." The rendering of this, according to the significance of this word, would be this: "And the fourth river is that one which makes the garden fruitful; that is, in which the garden yields its fruit."

Summing up the signification of the several rivers and countries, we have, first, the river that is the extension of the mouth, which, changing the character of its waters as it flows, encircles the whole of that which suffers pain and brings forth; second, a river that bursts forth from the valley of grace, which is in darkness, and where there is heat; third, a river that runs with a swift current and a sharp sound to the light, in front of the happy land; and fourth, a river that makes the garden fruitful.

The meaning of this summary is too evident to be escaped. The signification of these rivers is descriptive of the functions and of the various physical facts and capacities of the garden: they inform us how that garden is fed with new, and how drained of refuse or old and worn-out matter; they set forth the method by which the garden is made productive. Can there be anything more added to point the application with greater directness and force, save to designate the garden by the name by which it is now commonly known?

This Garden of Eden is a very much despised place; and if I were not to prepare the way, and guard every word I utter about it with the most scrupulous care, some of you might be so very innocent (by innocence, you must know, I mean that kind which comes of ignorance), or so modest (by modesty, you must know, I mean that kind which is born of conscious corruption, and which blushes at everything, and thus unwittingly proclaims its own shame)—I repeat that, if I were to approach the culmination too abruptly, such innocence and such modesty as that of which I speak, should there happen to be any present, might be too severely shocked.

At the outset, I must ask you to remember that it is out of the most despised spots of the earth that the greatest blessings spring; that it is out of the most obnoxious truths that the forces are developed which move the people heavenward fastest. It is the same old question, "Can there any good thing come out of Nazareth?" It should also be remembered that Jesus was conceived at the most despised of all the places of Galilee. The Jews could not believe that a Saviour of any kind could come from such a source. The promulgators of the new truths have ever been, and probably ever will be, Nazarenes; that is, will be the despised people of the world—though the meaning of that term in the original language is, "consecrated or set apart." It was in this sense that Jesus was a Nazarene. It was in this sense that the prophets were able to foretell that he would be a Nazarene. They knew that he would be set apart to do the greatest work of the ages, and therefore that, at first, he would be despised by the great of this world. Therefore, when we shall find the Garden of Eden, we may expect that it will be among the most despised, ignored, and ostracized of all the despised things of the world.

Lo, here—or, Lo, there—is Christ! is the cry of the world, which is always looking in the wrong direction for Him. Jesus said, "The Kingdom of God is within you." Suppose we find that the Garden of Eden is also within you? If the human body be a place worthy to be, and indeed is, the Kingdom of

God, it cannot be sacrilegious to say that it is also worthy to be, or to contain, the Garden of Eden. There cannot be a more holy place than the Kingdom of God; although I am well aware that too many of us have made our bodies most unholy places. Paul said, "*Know ye not that ye are the Temple of God; and that the Spirit of God dwelleth in you? If any man defile the Temple of God, him will God destroy.*" Then, the human body is not only the Kingdom of God, it is the Temple of God. Suppose, I say again, it should, after all, turn out that the long-lost Garden of Eden is the human body; that these three, the Kingdom of God, the Temple of God, and the Garden of Eden, are synonymous terms and mean the same thing—are the human body? Suppose this, I say. What then? Would not the people be likely to regard it with a little more reverence than they do now? —and to treat it with a little more care? Would they not modify their pretences that, in their natural condition, any of the parts of the body can be vulgar and impure, and unfit to be discussed either in the public press or the public rostrum? Is it not fair to conclude that, with a higher conception of the body, this ought to be the result? Certainly it would be, unless the doctrine of total depravity is true in its literal sense.

I am well aware that there must be a great change in the present thoughts and ideas about the body before it can be expected that there will be any considerable difference in its general treatment. But a great change has to come, and will come. Certain parts of the body—indeed, its most important parts—are held to be so vulgar and indecent that they have been made the subject of penal laws. Nobody can speak about them without somebody imagining himself or herself to be shocked. Now, all this is very absurd, foolish, and ridiculous, since, do you not know, that this vulgarity and obscenity are not in the body, but in the associated idea in the minds of the people who make the pretence; especially in those who urge the making of, and who make these laws, and who act so foolishly as to discover their own vulgarity and obscenity to the world in this way. How long will it be before the people will begin to comprehend that Paul spoke the truth when he said, "To the pure all things are pure." He ought to be good authority to most of you, who profess him so loudly. But I must confess that I have yet to find the first professing Christian who believes a single word of that most truthful saying. I fear that the hearts of such Christians are still far away from Jesus. But give heed to the truths to which I shall call your attention, and they will help to bring you all nearer to Him both in lip and in heart.

The despised parts of the body are to become what Jesus was, the Saviour conceived at Nazareth. The despised body, and not the honoured soul, must be the stone cut out of the mountain that shall be the head of the corner, though now rejected by the builders. There can be no undefiled or unpolluted temple of God that is not built upon this corner-stone, perfectly. And until the temple shall be perfect there can be no perfect exercise by the in-dwelling spirit. "The stone which the builders disallowed, the same is made the head of the corner."—1 Peter ii. 7. Christians have been thinking of taking care of the soul by sending it to heaven, while the body has been left to take care of itself and sink to hell, dragging its tenant with it.

"That through death he might destroy him that had the power of death, that is, the devil."—Heb. ii. 14.

"And deliver them who through fear of death were all their lifetime subject to bondage."—Heb. ii. 15.

"God hath chosen the foolish things of the world, to confound the wise; and God hath chosen the weak things of the world to confound the things which are mighty; and base things of the world, and things which are despised, hath God chosen."—1 Cor. i. 27, 28.

"And those members of the body, which we think to be less honourable, upon them we bestow more abundant honour."—1 Cor. xii. 23.

The last two chapters of the Revelation refer to the human body saved, and as being the dwelling place of God. The first two chapters of Genesis refer to the body, cursed by the acts of

31

primitive man (male and female), through which acts they became ashamed and covered themselves, because they had done evil to the parts that they desired to hide. Remember, that to the pure all things are pure ; and do not deceive yourselves by believing that anything which can be said about the natural functions and organs of the body can be otherwise than pure. From Genesis to the Revelation the human body is the chief subject that is considered—is the temple of God, which through long ages He has been creating to become, finally, His abiding place, when men and women shall come to love Him as He has commanded that they should; and this important thing is the basis of all revelation and all prophecy.

The objection that will be raised against accepting the evident meaning of the 2nd and 3rd chapters of Genesis will be that the things of which they really treat could never have been the subject of scriptural consideration. The degradation of the human race, following the transgression of Adam and Eve, through which purity was veiled from their own lustful gaze, and virtue shut out of the human heart, can never be removed until the world can bear to have that veil lifted, and to look upon and talk in purity about the whole body alike. It was not because they ought to have been ashamed of the nakedness they desired to hide, but because their thoughts were not pure and holy, and because their eyes could not endure the sight without engendering lust within them. So it is now. Only those are ashamed of any parts of the body whose secret thoughts are impure, and whose acts represent their thoughts whenever opportunities present themselves, or can be made.

People talk of purity without the least conception of the real meaning of the term. The people who do no evil because they have no desire to do it, are infinitely more virtuous than are they who refrain because there is a legal or any other kind of penalty attached thereto. So it is with the relations of the sexes. They are the really pure who need no law to compel them to do the right. I do not say that the law has not been useful, nor that it is not useful still. It is better to be restrained by law from doing wrong, than not to be restrained at all; but it is those who need restraint who ought to be ashamed, and not those who have grown beyond the need of law and wish for freedom from its force. In one sense, as Paul said it was, "the law was our schoolmaster;" but those who have graduated from the school, no longer need a master. Shall they, however, be compelled to have one, merely because all others have not yet graduated? Shall everybody be compelled to stay at school till everybody else has left? Think of these questions with but a grain of common sense, and you will see that they who urge the repeal of law are the best entitled to be considered pure at heart, as well as pure in act.

Jesus said, that "Whosoever looketh on a woman to lust after her, hath committed adultery with her already in his heart." Judged by this standard of purity, who are not adulterers? I will tell you who, and who only. Only those are not who can stand the test of natural virtue; and this test is never to do an act for which, under any circumstances, there is cause to be ashamed. Adam and Eve were not ashamed until they had eaten the forbidden fruit—the fruit of the tree which stood in "the midst of the garden," "whose seed is within itself;" but the moment they had done what they knew to be a wrong, when they had learned of good by knowing evil as its contrast, by reason of having done the evil, then they were ashamed and made covers for themselves. They are sexually pure and virtuous who enter into the most sacred and intimate relations of life just as they would go before their God, and by being drawn to them by the Spirit of God, which is ever present in His temple.

This is to have natural virtue. This is to have natural, in place of artificial purity. People who are pure and virtuous may be brought into intimate relations, and never have a lustful thought come into their souls. Now, this is the kind of virtue, purity, and morality that I would have established; it is the kind I advocate as the highest condition to which the race can rise. Suppose that the world were in the condition in which I speak, do you not know that it would be a thousand times more pure than it is? But do you say that all this is too far in the future to be of any use now? This plea is often made—that it ought not to be given to the people till they are ready to receive it and live it. I cannot have a more complete endorsement than to have it said that the people are not yet good enough to live the doctrines that I teach. But if they really do imagine this, I can assure them that they do not give the people credit enough for goodness. Bad as they are, they are not half so bad as some would make them out to be. Place men and women on their honour. You are all familiar with this principle, but you never think of applying it to the social relations, while it is really more applicable to them than it is to almost anything else. But, if the people are not good enough to live under the law of **individual honour**, then it is quite time that some one should have the courage to go before the world and begin to advocate the things that are needed to make them so.

Before leaving this part of my subject, I wish again to impress it upon you that when there is purity in the heart, it cannot be obscene to consider the natural functions of any part of the body, whether male or female. I am aware that this is a terrible truth to tell to the world, but it is a truth that the world needs to be told; one which it must fully realize before the people will give that care and attention to their creative functions which must precede the building up of a perfected humanity. Who shall dare say that the noblest works—nay, this holy temple—the kingdom of God—is obscene? Perish the vulgarity that makes such thoughts possible.

Where should the Garden of Eden be found if not within the human body? Is there any other place or thing in the universe more worthy to be called an "Eden"? Then let who may, esteeming himself a better judge than myself, condemn this garden as impure. If the gravity and grandeur of this subject were once realized you would never think meanly of, or desecrate your own body, but instead, you would do what Paul commanded (1 Corinthians vi. 20): "Glorify God in your body."

Anyone who will read the second chapter of Genesis, divorced from the idea that it relates to a spot of ground anywhere on the face of the earth, must, it seems to me, come to, or near, the truth. I have shown, conclusively, that it is not a garden in the common acceptance of that term: indeed, that the Garden of Eden, according to Moses, is a physical absurdity, if it be interpreted to mean what it is held to mean by the Christian world.

The Garden of Eden is the human body; the second chapter of Genesis was written by Moses to mean the body; it cannot mean anything else. Furthermore, Moses chose the language used because it describes the functions and uses of the body better than any other that he could choose without using the plain terms. Could there have been a more poetic statement of what really does occur? What more complete idea could there be formed of Paradise than a perfect human body—such as there must have been before there had been corruption and degradation in the relation of the sexes? *"Know ye not that ye are the temple of God, and that the Spirit of God dwelleth in you? If any man defile the temple of God, him shall God destroy, for the temple of God is holy, which temple ye are."*—(1 Cor. iii. ver. 16, 17.) *"What! Know you not that your body is the temple of the Holy Ghost which is in you? Therefore glorify God in your body."*—(1 Cor. vi. ver. 19, 20.)

But now let us go on with the application of our former inquiries into this garden: "And a river went out of Eden to water the garden, and from thence it was parted, and became into four heads. The name of the first river is Pison, as we have seen. It will be remembered that this term signifies changing and extension of the mouth. Now, apply this rendering to the body and see if we cannot find the river Pison in this Havilah, which we failed to find in the Arabian land. How is the body watered and fed? Is it not by a stream which is the extension of the mouth, and that changes constantly as it encircles the system? Does not the support of the body enter it by the mouth, and by the river which is the extension of the mouth run to the stomach? "And from thence it was parted, and became into four heads." Now this is precisely what is going on in the body all the time. From the stomach, or rather from the small intestines, where the separating process in the chyle, which is the digested contents of the stomach, begins, this river Pison has four principal heads; that is, it divides and becomes into four heads, giving off three branches, while the main current continues on its course to compass the whole land of Havilah. This current—this river Pison—empties itself into the heart, and then into the lungs, where it is de-carbonized and oxygenized, and returned to the heart to be distributed over the entire system by the arterial circulation. In its course toward the extremities it gives to the various parts through which it passes their necessary supplies. This constant giving-off changes the character of the current as constantly, until the circumference of the body is reached. From thence it is returned to the heart through the venous circulation, gathering up the worn-out matter to expel it from the body. This is the process by which the river Pison compasseth the whole land of Havilah, which is the land "that suffers pain and brings forth," and in which there are precious things, besides the bdellium and the onyx stone. This land that suffers pain and brings forth is the land of Havilah, which is compassed by the river Pison. Can anyone conceive a more graphic description of the process by which the body is nourished and fed? A river, to water the land of pleasure or delight, enters by the mouth, and extending by the way of the stomach, intestines, heart, lungs, arteries, and veins, waters the whole land that suffers pain and brings forth. What is there in the world to which this description of the river Pison and the land of Havilah could be applied, save to the body? It cannot be found. I challenge the world to find it. It would be absurd, simply, to say that the district south-east of Sanaa, in Arabia, which is called Havilah, suffers pain in bringing forth. Nevertheless, this is the conventionally accepted land of Havilah.

"And the name of the second river is Gihon: the same is it that compasseth the whole land of Ethiopia." The first branch that divides from the main river of the body is that which drains the body by way of the intestines. This is the river Gihon, which is the valley of grace. Could there be a more appropriate name than that of "grace" for the process by which the refuse from the river Pison is discharged from the body? or than the valley of grace for the operations that are performed within the abdomen for the elimination from the body of the refuse that is gathered there. Is not this a process of grace?—a process of natural and involuntary purification? If it were not for this purpose of grace we should be lost through the *débris* of which the system is relieved by this bursting forth of the river Gihon from this valley of grace.

And this is the river that compasseth the whole land of Ethiopia—the land of blackness (darkness), and where there is heat (see Psalm cxxxix. 12). That is to say, the intestines occupy the abdominal cavity, which is the land of darkness in Eden. All the movements that are made therein are made in darkness, and therein also is the heat which signifies the warmth that gives and maintains life; that maintains the old and that produces the new; that sustains the temperature of the body, and that gives it the power to reproduce. Physiologically this is absolutely true, just as are all the other descriptions and allegories that are given by Moses of the garden.

"And the name of the third river is Hiddekel, that is it

which goeth toward the east of Assyria." Next in importance to the maintenance of the human economy is the river that drains the system of another class of impurities, running by the way of the kidneys, uretus, bladder, and urethra. This is the river Hiddekel; or the stream that runs with a "swift current" and a "sharp sound." Search the language through and through for a more appropriate description for the elimination of the waste matter by the means of the urinary organs than this one given by Moses. And this river of Eden runs toward the east of Assyria, which is the "land of the garden," in the midst of which is the tree of life. That this may be still more evident, it is proper to remark here, that it is the female human body which is referred to by Moses, because it is her body that suffers pain in bringing forth; and it was the producing part of the garden—the reproductive female power—that was the land which was cursed in Eden by the transgressions, *by eating of the fruit of the tree of life improperly*. It was by this curse that woman's "*sorrows and conceptions were multiplied*," as stated by Moses. So the Garden of Eden is the producing land of the human family into which the Lord God put the man whom he had formed, "to keep it and to dress it," so that it might be fruitful. Do you not see how perfect the allegorical statement is, which Moses made?

"And the fourth river is Euphrates." The last river of the Garden of Eden is that one which renders it fruitful; that makes it yield its fruit, and that flows through the reproductive system. Euphrates means fruitfulness, and this river, the last one in the order of physiological sequence, is the fruit or the result of the perfected action of all the others combined. This river was in its natural, healthful, primitive state of purity, from which physical purity primitive man and woman fell by the improper use of the functions of the garden, which were committed to their care, the same as people continue to do, and are cursed—die in Adam.

At the time when knowledge began to find root in the brain of man, it is pretty evident that the human animal, man, was pure and perfect physically; that is, that they were like the other animals, and that they are to be judged of as we judge of animals now. Considered in this light, what are the differences between man and the animals? This is a question of the most vital importance, since, if there was a fall of man from the original state of purity, it is necessary that we know of what that fall consisted before we can provide intelligently for an escape therefrom. It was not a moral fall certainly, since morality is not an attribute of animals, unless physical purity is morality. This view of ethics is not legitimate, since morals are the last development in the growth of man, are an outgrowth of, or a building upon, intellect. Nor could that fall have been intellectual, since as there had then been no knowledge of good and evil, there was no intellect; there had been no power of comparison in the human brain. We are obliged to conclude, therefore, that that sin committed by man was a physical sin.

Now what was this sin? Well, go to the animal world and compare its physical habits with our own, and it will not be difficult to discover differences sufficient to account for all that has occurred to mankind; indeed, we shall find such a disparity that we shall be left only to wonder that a second deluge has been so long deferred. What is the central point towards which all these differences gravitate? It is clearly the relations between the male and female. Undoubtedly, before the fall of man, if we accept the Biblical story, these relations between the sexes were the same then, as they are now, between the animals; that is, they were solely for propagation, and in this respect the female was and is supreme mistress.

But what has been the result of this desecration of woman? Look again to the female animals and learn; for here woman stands in lurid contrast to her sex in that domain! Where is the animal that wastes her very life at every changing moon? There are no such, except among the monkeys, and the fact exists there for the same reason that it exists among women.

Menstruation is an hæmorrhage or exfoliation of the mucous membrane of the uterus. Some of the higher mammals have something similar at the period of œstus or heat, but monkeys are the only mammals which menstruate like women. Monkeys are the only mammals which copulate for other purposes than reproduction.

Apes were kept in confinement thousands of years before King Solomon's ships brought home from Tarshish "ivory, apes and peacocks" (2 Chron. ix., 21). A monkey (kaf) appears under the chair of a person who lived in the reign of Cheops, 4th dynasty, proving that the word is much older than the Sanskrit form (Wilkinson, Ancient Egyptians, vol. iii., p. 269: See also vol. ii., p. 190). Aristotle, who wrote upon the resemblance of man to the monkey, takes no note of this fact; "some animals unite in their natures the characteristics of man and quadrupeds, as apes, monkeys, &c."

Pliny in his "Natural History" speaks of human beings and of monkeys, but of the things *wherein they are alike* no mention was made. Was Pliny ignorant of this fact? Monkeys and women are the only animals which menstruate. They are the only mammals which copulate promiscuously and at every season. Monkeys are still monkeys in spite of this fact; they have not evolved to something higher, so neither menstruation nor the increased excitation of the generative organs could have been the cause of subsequent development.

Pflüger has said "that menstruation is the result of the growing follicle acting as an irritant to the terminations of the nerve fibres embedded in the stromas of the ovary. This irritation finally brings about congestion of the genital organs by an afflux of blood to these organs. Other animals besides monkeys and women have ovaries and yet the pressure of the growing follicles does not cause congestion of the genital organs out of season. Moreover, if menstruation were analogous to the æstus of animals, this only appears at certain periods. The sum of the irritations of the growing Graafian follicles is not the same, evidently, as in monkeys and women. But I might ask why should the sum of these irritations become so great at periodic intervals causing the periodic congestion of the genital organs? When the ovaries become atrophied and menstruation ceases the woman is no longer capable of producing life; hence the expression, turn of life. Upon the maturity of the Graafian follicle menstruation commences, and it indicates that the female is able to produce life. Menstruation itself is not essential to life as animals who never menstruate produce life, and girls have become pregnant who have never menstruated. The essential principle of life is seated in the ovaries, and it is the excitation out of proper seasons which has brought on menstruation in the female organism. We say we cannot re-pot or transfer plants if it is not the proper season, we must not disturb the roots while the sap is running. If the ovaries are cut out of a girl before she has menstruated she never does so. Thousands of years ago the ancients must have arrived at some idea of the truth when they caused the ovaries to become atrophied by puncturing with needles which had been dipped in chemicals. Did this custom have its origin in some attempt to solve the phenomena of life?

The ovaries are not only essential for the function of menstruation, but they are also essential for the development of the female generative organs. If the ovaries are degenerate, atrophied, or arrested in their growth, the pelvis remains narrow, the uterus, the breasts, &c., are undeveloped. In diseased ovaries menstruation is intermittent or ceases; in diseased ovaries the secondary sexual characters are apt to reproduce themselves. The ovaries are responsible then for menstruation, its cessation, and the development of the generative organs. Tilt in his book on Ovarian and Uterine Inflammation, says, "when there is no ovary the uterus, should it exist, does not menstruate. It is the ovary which calls the uterus into action, imparting to it a stimulus which is either healthy or morbid, periodical or continuous. . . Menstruation is a species of parturition. *The reproductive organs are indeed the only organs of the body whose function is painful even when healthily performed."* . . . Dr. Tyler Smith sought to prove that the bulk of diseases of women originate in the *hyper-secretion* of the mucous glands of the neck of the womb. . . . Dr. Ashwell says, Of all the organs of the human body scarcely any seem so prone either to functional or organic disuse as the ovaries; for I can with truth say that I have rarely when examining these important organs after *death found* them entirely *healthy*. . . . In Germany, Neumann did not scruple to remark that, of all the organs of the human frame, none are so often affected by disease as the ovaries. . . . If menstruation does not take place when the ovaries are absent, it follows that menstruation had its origin in something which affected or related to the ovaries. As no mammal menstruates where coition takes place for reproduction only, it was the copulation for other purposes than the perpetuation of the species which brought on menstruation—the undue excitation of the ovaries. The fact that menstruation ceases when the fertilized ovum becomes attached to the uterus would seem to prove that menstruation is the hereditary result of the *excitation of the ovaries* for other purposes than for propagation, for if there are no ovaries there is no menstruation.

It may be asked, what stimulated monkeys to copulate for other purposes than propagation? When the transition took place from the attitude of the quadruped to that of the biped, the pelvic viscera, by impact, were pressed down towards the pelvic outlet. Undue secretion of the mucous glands of the vagina corresponding to the æstus or rut of animals may have been the result. This may have incited the males to rape the females.

This truth was realized by the most ancient religious sects. They found it necessary to check the superior brute force of the male from desecrating the female by promulgating such religious laws as these. "The birth of a child is a defilement to its parents, especially to its mother who is declared impure for as many days as have elapsed months since her conception, and her purification shall be accomplished as after her natural seasons." (Manou in the Vedas.) In Leviticus we read that "her purification shall require sixty days." In the Veda—"the husband should respect his wife in her natural seasons as we respect the blossom of the banana which announces fecundity and future harvest." If we study different religions we find it has been necessary to make religious laws to protect woman from violation. There stands the fact, ye women of the world, and there is where ye differ from the animals; and in this fact all the results of the original sin have had their source. Let any female brute lose the control of her procreative functions, as woman has been deprived of hers, and let her be subjected to the unbridled passion of the male, and she will soon begin to menstruate.

No animal menstruates which copulates for reproduction only. Some scientists have suggested to me that this may have been the cause of the subsequent intellectual development of the human race. I see no scientific truth in this theory.

Idiots have their generative organs abnormally developed. There is a great scientific truth in this which I am investigating for a future work. Except in morbid pathological conditions, the cerebral soul developes in degree as it overcomes the abdominal soul. The licentious monkeys and savages with their thick protruding lips, indicating great sensuality and small mental capacity would negative this hypothesis were there no other refutation. But if it needed any further refutation I have only to give an example from history.

History is a great teacher—it enables us to learn by the experience of others. The Spartans were taught self-control, simplicity in their way of living, that their actions must be directed by Reason, and that the vital law manifested through their beings was not to be made the instruments of beasts but the creator of gods. Here licence was not accorded, but restraint was enforced. What was the result? The Spartans were known as the Invincibles, powerful both in body and mind. History teaches us another lesson. At Rome in the days of Juvenal when there was no such thing as self-control; Reason was a myth, civilization a mockery, and purity a jest. Energy directed too much in one direction must be at the expense of some other part. There must be an equivalence. We cut back the leaves of a plant when we want abundant bloom and pick off the bloom when we desire foliage.

An athlete who expends his energy in muscular exertion has not that energy left to expend in the artistic perception and muscular work of the eyes, or the musician of his ears. There is in either case mechanical work done and energy expended, but one is at the expense of the other. The energy of the human body is limited. If we use up our energy in diseased appetites, we have not that energy left for noble thought and artistic pursuits. The brain has developed by the *exercise* of the reasoning faculties, more exact or cumulative methods of observation. Every new scientific discovery is the means of adding to our knowledge. We have an example of what education and training can do with the youthful mind in the great disparity between men and women in this regard—the systematic collegiate education of boys and the makeshift education which has been in vogue heretofore with regard to girls.

We still further see the result of this sin in all those false ideas which are being disseminated with regard to preventive checks or means to prevent conception, as the cure for bad population. We are told that the amative impulse may be allowed full scope, so long as children be not produced, save as and when desired. Science is cited as the instrument which enables us to have many domestic comforts—that the aid of science is called where it would be injurious to the mother to have a child—why not under other circumstances? These books are criminal in their ignorance of natural laws.

Menstruation being the hereditary result of the undue excitation of the ovaries, there could be no greater condemnation for those who advocate preventive checks.

Morbid menstruation or excessive excitation of the ovaries induced by sexual excitement is the cause of ovarian and uterine inflammation, of uterine tumours, of ovarian cysts, of ovarian dropsy, of cancer of the cervix uteri, and various other pathological conditions of the generative organs. And this pathological condition is by no means confined to the individual life. It may be handed down from generation to generation for diseased ovaries are found in mere children; cases are on record where ovariotomy has been performed on young girls, one only eight years of age, for ovarian tumour. Cysts are found in the ovaries of new-born children, showing that these pathological conditions are hereditary. And yet these ovaries influence the whole body, often rendering existence one lifelong martyrdom. Poor mortals doomed to a living death! Degenerate organs of reproduction produce monsters of every description. The enormous percentage of women who die from these causes can only be realized by reading medical books on diseases of women. Constant irritation of the ovaries reacts upon the nervous centres, producing all kinds of morbid effects, headaches, lassitude, irritability, nervous chills, hysteria, hypochondria, melancholia, epilepsy, paralysis, hyperexcitability, lethargy, catalepsy, somnambulism, strange alienation, and various degrees of insanity. It is not the frequent child-bearing which is so disastrous to the mother, but the constant drain upon her available energy by too frequent sexual excitement and consequent exhaustion.

Why is it that when the testes are extirpated in the male the secondary sexual characters do not produce themselves? And when the ovaries are extirpated in the female, the secondary sexual characters do produce themselves? This would show that the test react upon the body of the male to cause increased vital activity. It is just the opposite with the female, though with savages and in some parts of the East where ovariotomy is performed for certain purposes, it is said that the women develop unusual strength. In the female the ovaries drain to themselves and store up energy for the future embryo, and when the ovaries become atrophied or are extirpated, the energy not being needed for the ova reacts upon the whole organism.

Ovarian activity therefore always means a loss to the female, a drain upon her available energy. The female organism being called upon to produce aborted life in the form of the monthly or frequent maturation of the Graafian follicle, the false corpus luteus of menstruation is a drain upon and waste to the maternal organism and no benefit to the race. Woman, as she is to-day, can be said to be undergoing perpetual childbearing, her generative organs are never at rest, except before puberty, and after she has ceased to be fruitful. The hypothesis has been advanced, that longevity or length of life is the adaptation to the needs of the offspring, that where animals deposit a large number of eggs, there is more likelihood of some surviving, therefore the maternal organism dies young, in some instances death follows immediately. In those cases where there are a small number of offspring, or the number is limited, there is a tendency for the life of the maternal organism to be prolonged to secure the perpetuation of the species.

If this hypothesis is true, the frequent maturation of the Graafian follicle is so much loss of vitality to the human race. In those races where puberty of the female is retarded she retains her youth longer, and the average duration of the life of the race is the longest. And in those countries where the sexual sentiments have been worked by the custom of early marriages, the hereditary result has been early menstruation, shorter life, and deterioration of the race. The vital principle is developed and determined by the female and is followed by the male. Even the greater developed mental capacity of the male is largely dependent upon the mother who bore him.

In repeated pregnancies the weight of the child and the bulk of its head are increased. Schroeder, in his "Manual of Midwifery," says: "The weight of the child increases with the age and especially with the number of the previous labours of the woman. ... The heads of male children are larger than those of female, and the most important diameter of the head—and the biparietal—increases quite out of proportion to the number of labours and the age of the mother, so that the broadest skull may be expected in a male fœtus of a pluriparæ somewhat advanced in age . . . . consequently not to expect very large heads in young primiparæ, whilst in an older woman who has often borne children, a head of a considerable size may be looked for. This proves that the function of gestating is developed in successive pregnancies, that the fœtus is better nourished, and has had the advantages of the more fully developed maternal organs. But our civilization would provide means to avoid conception when the mother has had one or two children, so that those children, who would be more developed physically and mentally, should not be born, or if born in spite of preventive checks should have the injurious effects resulting to contend against.

Sir Spencer Wells, in his book on "Ovarian and Uterine Tumours," published in 1882, gives a brief sketch of the history of Ovariotomy, and quotes the following:—

"A paper was laid before a late meeting of the Anthropological Society of Berlin for publication in their Transactions which reports that the aborigines of Australia and

New Zealand performed ovariotomy on young girls (the age is not mentioned) by incision in both inguinal regions. They do this for two purposes: first to *prevent the propagation of hereditary diseases and deformities, and other disabilities.* The writer met a woman born deaf and dumb who had been spayed to hinder her from bearing deaf and dumb children. . . . For the same reason of personal defect men are made impotent."

The preceding indicates that certain savages have at least some regard for the future members of their community. It is to be hoped, however, that when the responsibilities of parentage become more fully recognized that neither method —preventives nor spaying, will be necessary; the force of educated public opinion will deter the unfit from propagating their kind: but if human beings have not sufficiently evolved to realize that the function of their generative organs is reproduction, I would recommend the remedy of the ancient Greeks.

Scientific propagation is not a new idea which has originated with our nineteenth century civilization.

Menstruation is the result of the excitation of the ovaries, and the function of the ovaries is to develop and bring to maturity the Graafian follicle; and, as function has always preceded structure, we must seek the cause of the deviation of structure in the pathological condition of the ovaries—inflammation, cystic degeneration, dropsy, &c. The same may be said of the changes which the fertilized ovum may undergo in the uterus, or why is the purity of an animal's blood lost when it has once been crossed? It cannot be that the developing fœtus can affect a future ovum which is not yet mature. It is because the function of gestation is affected, and, if so, what effects uterine inflammation, tumours, cancer, &c., must have! What functional derangements may not be given to the developing fœtus? This negatives the hypothesis that differentiation of species is the result of the segmenting female and male pronucleus, and that only. When the generative organs are diseased they react upon the nerves governing nutrition, and this is of vital importance to the pregnant woman, for how can she impart to her child energy and rich blood when there is hyperæsthesia or anæsthesia of the nervous centres affecting these organs? And if psychical processes are to be traced back to physical processes the mother moulds the character of her child.

If we sought for the cause of nine-tenths of the insanity which is the curse of the nineteenth century, I think we should find it in sexual debauchery; that those insane had inherited weakened nervous systems through the sexual debauchery of their ancestors, or abuse in childhood of their own generative organs, or debauchery in maturity. Only the doctrine that under no circumstances ought humanity to propagate unless healthy both in body and mind, will perfect the human race. The loss of the prerogative of becoming creators when they sin against the God who gave them the right to be, would be an incentive to our sons and daughters to study such laws of life as will produce physical and psychical perfection by appealing to one of the most powerful instincts.

The maternal and paternal instincts to propagate are natural instincts; but these instincts have become diseased, and this disease is known by the term lust.

Is it not to be wondered rather that the human race enjoys anything that can be called health? Aye, still more, is it not astonishing that it even lives at all—that it has not long since been swept from the face of the earth, as it soon will be if it do not repent of this sin?

All this is feminine, since it is from the waters of the river Euphrates that the fruit of the tree of life (whose seed, as Moses said, is within itself) is developed and perfected. But this stream of life was turned to blood by the transgressions of primitive man, and has been entirely wasted to the race save that small portion which is utilized during gestation. The supposition that this river is something of which the female system ought to be relieved—that it is lifeless and corrupt—is false and wrong.

If it had no physiological value, why does menstruation cease during the period of gestation? Nature does not deem this blood corrupt and valueless then. From the Veda of ancient India is the following: "The blood is the life, it is the divine fluid that waters and fecundates the matter of which is formed the body. It is through the blood that the pure essence emanating from the Great Whole, and which is the *soul*, unites itself to the body."

But this river of life has been left to waste away the health and strength—the vigour and vitality—of the race, and no efforts have been made to remedy the destruction which it threatens, a disaster involving the fruitfulness of the garden itself, and the consequent wiping out of the race. This wasteful process is considered to be a natural function, and necessary to health and life, and so indeed it is, in the unnatural conditions in which we live, and in which the world has lived since this river was turned to waste, as described by Moses, allegorically, in the 4th and 7th chapters of Exodus. This wasting away of the life of the race is the vicarious atonement by which death is averted for the time. The fulness of time is not yet; the race, for a time, must rush madly onward toward destruction and extinction; but when the New Jerusalem (which is the purified woman) shall come in the new heaven and the new earth, as seen by St. John on Patmos, then this river of waste will return again to be "a pure river of water of life proceeding out of the throne of God"—proceeding out of His highest creative place, the Garden of Eden, through which flows the fruitful river Euphrates.

The Garden of Eden then is the human body, and its four rivers, which have their source in the extension of the mouth,

50

are the Pison, the blood; the Gihon, the bowels; the Hiddekel, the urinary organs; and the Euphrates, the reproductive functions. By these four rivers the whole garden is watered and fed or nourished and supported, drained of refuse matter, and its fruit produced. It was in this garden that mankind was planted by the Lord God after the same manner in which He performs all His other works—through the agency of law and order, as exemplified in evolution. It was the ground of this garden that was cursed, so that in sorrow man should "eat of it all the days of his life," and that it should bring forth "thorns and thistles," as Moses said it should, instead of the pleasant and agreeable fruit of perfect and beautiful children. Has not this allegorical picture been literally verified? Paul said he had only "the first fruits of the spirit;" that is to say, having the intellectual comprehension of the means for redemption of his body only.

"If any man defile the temple, him shall God destroy." Does He not do this? Does not death follow the defilement of the temple? In the temples that man has erected, and into which he enters on every seventh day to worship God, He does not dwell. These are the figures or the images only, as Paul said, of the true temple. Neither in this mountain nor at Jerusalem shall man worship; but in spirit and in truth, said Jesus. The fact that there are so many temples made with hands, into which all the professedly Christian world feels it to be necessary to enter and worship, is a certain evidence that their temples, not made with hands, are not yet the abode of God. Not having consciously the kingdom of heaven within them, where God comes and dwells with them, they still go after Him; and they are so blind that they do not see their own condemnation in the act. If a person has God dwelling in him, he need not go to church to worship Him, nor by so doing to make it evident to others that he is one of God's people, to whom He has come, and with whom He has taken up His abode.

Those who have to make a profession of faith to make it appear that they have God, only expose their own hypocrisy, for God's presence in any human being is self-evident proof of the fact.

Consider for a moment what would be the result if the people could come to recognize that their bodies are God's holy temples, and that their sexual organs, being the means by which His crowning work is created, ought never to be defiled by an unholy touch or thought, or ever made the instruments of selfish gratification merely. If the people should enter into these sacred relations only as if they were communing with God—with the same spirit in which really earnest and honest Christians enter into the temples made with hands, which they have falsely thought to be God's temples—and not with unbridled passion, what would become of the debauchery that now runs riot in the world? No; let the sexual act become the holiest act of life, and then the world will begin to be regenerated, and not before. Suppose that those who read the Scriptures, and pray regularly before eating, should go through the same ceremony before entering into the relations which should be the holiest of all relations, how long would the beastliness that now holds high carnival under cover of the law continue? If praying people believe the Bible—believe that their bodies are God's temple—why should they make such hot haste to defile them by their selfish lust and inordinate lasciviousness? Let these people become consistent at least, and in the most important act of life ask God's blessing to rest upon it.

But John saw that these fruits were to be fully realized in the new heaven and the new earth, meaning the new man and the new woman.

"Adorned as a bride prepared for her husband." Can there be any mistaking the significance of this figure? Can it mean anything save the perfected union of the sexes; and in the understanding that this perfection is coming to the world?

Why has God permitted His people to live in darkness and

death (all die in Adam) so long, the Christian will ask; and if there is any truth in the Bible as being God's truth, why did He not make it so clear that none could misunderstand it and be lost thereby, the scientist will retort. Now, here is precisely where the reconciliation between religionists and scientists will come. The very thing that the Bible declares to be a gift of God, which is to be revealed when the mystery shall be solved, is the very thing after which all science seeks—the perfect life. The ultimate fact after which both religion and science bend their energies is the self-same thing. The Spirit—God—tells what this is inspirationally in the Bible; men delve for it among the laws of nature scientifically. At the same time that it shall be discovered to the world of what this mystery of God consists; it will be demonstrated by actual life in individuals. Inspiration and evolution mean the self-same thing, spoken from the opposite extremes of the development by which it shall come—the former being the spiritual comprehension of the truth before it is "made flesh and dwells among us," and evolution being its actualization in experience.

Interpret the arbitrary commands of the Bible by the language of natural law, by which alone God works, and the reconciliation between God and nature, between religion and science, between inspiration and evolution, is completed. Inspiration is the language of men who were permeated with Divine essence, but knew nothing about the law of cause and effect. They attributed the destruction of a city by fire or by an earthquake, in short, every visitation of painful effects upon men, as a direct and arbitrary command of God as punishment for sin; while by the light of science they are only the natural effects of immutable laws, occurring because they must occur, in the evolution of the universe. All the sins and punishments of which man has been made the subject are of the same order. It was impossible that man, being an animal, should be made a son of God, save by the very process through which he has had to pass.

53

That the law of evolution which makes growth the method by which intellectual altitude is reached, is also the law by which physical development goes forward; the perfected creation of man and his consequent salvation from death being physical and not moral, as has been falsely taught by almost the whole of Christendom. With a perfect physical body—man reconciled to God—all other perfections follow as its fruit, necessarily. The opposite proposition to this is the stumbling-block over which all Christians have fallen; they have given all their attention to saving the soul hereafter, when this salvation depends entirely upon saving the body here and now.

Is it not palpable how the acceptance of this fact, and the adoption of its logic as a rule of human action, would harmonize the relations of man? With this view, everything that occurs is a part, and a necessary part, of the evolution or the growth of man. *Suppose criminals were to be treated by this principle, what a reform might be inaugurated in this regard!* Suppose this precept were to be made a rule of life, the world could be at once transformed into a brotherhood. But this must also be a result of growth.

"And out of the ground (female-male) the Lord God formed every beast of the field, and every fowl of the air.' The two sexes must have been comprised in each species, evidently a rib was not taken out of each male to make a corresponding female.

In the first dawn of the life-principle there was no such thing as sex. Life was a unit, that is, a homogeneous mass, gradually becoming heterogeneous until two sexes were evolved. The Biblical allegory of Adam and Eve, that the two sexes were evolved from one, accords with science. Had this a deeper meaning than even Moses comprehended? Still more curious was the supposition that the male animal was the first distinct sex; before the male animal, it was the two sexes in one of the female-male animal. And the male organs of the latter becoming gradually degenerated or suppressed the distinct female animal was evolved to correspond with the male animal.

Here we have the ideal marriage. The two unite to become as one from which the human family had its birth. Onward from the family next were formed the roving tribes which had a chosen head, who ruled the whole with arbitrary will in all respects. Next cities sprang into existence, and reaching over provinces united into nations, making their kings or queens, their rulers absolute. From this, the concentrated form of power, the sway began to re-dispose itself among the people. Through monarchs limited in rule to constitutions and republics has the power descended and now it is about to be assumed again by each and all individuals who have become a law unto themselves, into whose hearts Almighty God has put His law of love. From individuals such as these a brotherhood of man can form and live, but not from any other kind. And from a brotherhood wherein the good of each becomes the good of all, the higher and the holier family will spring into existence, whose King and Queen and Lord and Prince shall be the living God who from creation's dawn through long experience, sometimes dark but often bright, hath brought us kindly on our way to this exalted place as His abode.

The Revelation of St. John the Divine.
Chapter XXI.

And I saw a new heaven and a new earth: for the first heaven and the first earth were passed away: and there was no more sea.

2 And I John saw the holy city, new Jerusalem, coming down from God out of heaven, prepared as a bride adorned for her husband.

3 And I heard a great voice out of heaven saying, Behold, the tabernacle of God *is* with men, and he will dwell with them, and they shall be his people, and God himself shall be with them *and* be their God.

4 And God shall wipe away all tears from their eyes; and there shall be no more death, neither sorrow, nor crying, neither shall there be any more pain: for the former things are passed away.

55

5 And he that sat upon the throne said, Behold, I make all things new. And he said unto me, Write; for these words are true and faithful.

6 And he said unto me, It is done. I am Alpha and Omega, the beginning and the end. I will give unto him that is athirst of the fountain of the water of life freely.

7 He that overcometh shall inherit all things; and I will be his God, and he shall be my son.

8 But the fearful and unbelieving, and the abominable, and murderers, and whoremongers, and sorcerers, and idolators, and all liars, shall have their part in the lake which burneth with fire and brimstone; which is the second death.

9 And there came unto me one of the seven angels which had the seven vials full of the seven last plagues, and talked with me, saying, Come hither, I will shew thee the bride, the Lamb's wife.

10 And he carried me away in the spirit to a great and high mountain, and showed me that great city, the holy Jerusalem, descending out of heaven from God.

11 Having the glory of God: and her light *was* like unto a stone most precious, even like a jasper stone, clear as crystal;

12 And had a wall great and high, *and* had twelve gates, and at the gates twelve angels, and names written thereon, which are *the names* of the twelve tribes of the children of Israel:

13 On the east three gates; on the north three gates; on the south three gates; and on the west three gates.

14 And the wall of that city had twelve foundations, and in them the names of the twelve apostles of the Lamb.

15 And he that talked with me had a golden reed to measure the city, and the gates thereof, and the wall thereof.

16 And the city lieth foursquare, and the length is as large as the breadth: and he measured the city with the reed, twelve thousand furlongs. The length and the breadth and the height of it are equal.

17 And he measured the wall thereof, an hundred *and* forty *and* four cubits, *according to* the measure of a man, that is, of the angel.

18 And the building of the wall of it was *of* jasper: and the city *was* pure gold, like unto clear glass.

19 And the foundation of the wall of the city *were* garnished with all manner of precious stones. The first foundation *was*

jasper; the second, sapphire; the third, a chalcedony; the fourth, an emerald.

20 The fifth, sardonyx; the sixth, sardius; the seventh, chrysolite; the eight, beryl; the ninth, a topaz; the tenth, a chrysoprasus; the eleventh, a jacinth; the twelfth, an amethyst.

21 And the twelve gates *were* twelve pearls; every several gate was of one pearl: and the street of the city *was* pure gold, as it were transparent glass.

22 And I saw no temple therein: for the Lord God Almighty and the Lamb are the temple of it.

23 And the city had no need of the sun, neither of the moon, to shine in it: for the glory of God did lighten it, and the Lamb *is* the light thereof.

24 And the nations of them which are saved shall walk in the light of it: and the kings of the earth do bring their glory and honour into it.

25 And the gates of it shall not be shut at all by day: for there shall be no night there.

26 And they shall bring the glory and honour of the nations into it.

27 And there shall in no wise enter into it any thing that defileth, neither *whatsoever* worketh abomination, or *maketh* a lie: but they which are written in the Lamb's book of life.

Chapter XXII.

And he shewed me a pure river of water of life, clear as crystal, proceeding out of the throne of God and of the Lamb.

2 In the midst of the street of it, and on either side of the river, *was there* the tree of life, which bare twelve *manner of* fruits, *and* yielded her the fruit every month: and the leaves of the tree *were* for the healing of the nations.

3 And there shall be no more curse: but the throne of God and of the Lamb shall be in it; and his servants shall serve them:

4 And they shall see his face: and his name *shall be* in their foreheads.

5 And there shall be no light there; and they need no candle, neither light of the sun; for the Lord God giveth them light: and they shall reign for ever and ever.

6 And he said unto me, These sayings *are* faithful and true: and the Lord God of the holy prophets sent his angel to show unto his servants the things which must shortly be done.

57

7 Behold I come quickly: blessed *is* he that keepeth the sayings of the prophecy of this book.

8 And I John saw these things, and heard *them*. And when I had heard and seen, I fell down to worship before the feet of the angel which shewed me these things.

9 Then saith he unto me, See *thou do it* not: for I am thy fellowservant, and of thy brethren the prophets, and of them which keep the sayings of this book: worship God.

10 And he saith unto me, Seal not the sayings of the prophecy of this book: for the time is at hand.

11 He that is unjust, let him be unjust still: and he which is filthy let him be filthy still: and he is that is righteous, let him be righteous still: and he that is holy, let him be holy still.

12 And, behold I come quickly; and my reward *is* with me to give every man according as his work shall be.

13 I am Alpha and Omega, the beginning and the end, the first and the last.

14 Blessed *are* they that do his commandments, that they may have right to the tree of life, and may enter in through the gates in the city.

15 For without *are* dogs, and sorcerers, and whoremongers, and murderers, and idolaters, and whosoever loveth and maketh a lie.

16 I Jesus have sent mine angel to testify unto you these things in the churches. I am the root and the offspring of David, *and* the bright and morning star.

17 And the Spirit and the bride say, Come. And let him that heareth say, Come. And let him that is athirst come. And whosoever will, let him take of the water of life freely.

18 For I testify unto every man that heareth the words of the prophecy of this book, If any man shall add unto these things, God shall add unto him the plagues that are written in this book.

19 And if any man shall take away from the words of the book of this prophecy, God shall take away his part out of the book of life, and out of the holy city, and *from* the things which are written in this book.

20 He which testifieth these things saith, Surely I come quickly. Amen. Even so, come, Lord Jesus.

21 The grace of our Lord Jesus Christ be with you all. Amen.

India, according to the Vedas, entertained a respect for woman amounting almost to worship.

MAXIMS FROM THE SACRED BOOKS OF INDIA.

"He who despises woman despises his mother."

"Who is cursed by a woman is cursed by God."

"The tears of a woman call down the fire of Heaven on those who make them flow."

"Evil to him who laughs at woman's sufferings: God shall laugh at his prayers."

"It was at the prayer of a woman that the Creator pardoned Man: cursed be he who forgets it."

"Who shall forget the sufferings of his mother at his birth shall be reborn in the body of an owl during three successive transmigrations."

"There is no crime more odious than to persecute woman."

"When women are honoured the Divinities are content; but when they are not honoured all undertakings fail."

"The households cursed by women to whom they have not rendered the homage due them find themselves weighed down with ruin, and destroyed, as if they had been struck by some secret power."

"The infinite and the boundless can alone comprehend the boundless and the infinite, God only can comprehend God."

"As the body is strengthened by muscles, the soul is fortified by virtue."

"The wrongs we inflict upon others follow us like our shadow."

"It is time to appreciate all things at their true value."

Let us repeat that story from Sufi: "There was a man, who for seven years, did every act of charity, and at the end of seven years he mounted the steps to the gate of Heaven and knocked. A voice cried, 'Who is there?' 'Thy servant, O Lord,' and the gate was shut. Seven other years he did every other good work, and again mounted the three steps to Heaven and knocked. The voice cried, 'Who is there?' He answered, 'Thy slave, O God,' and the gates were shut. Seven other years he did every good deed and again mounted the steps to Heaven, and the voice said, 'Who is there?' He replied, 'Thyself, O God,' and the gates wide open flew."

And why?

Because, as the Scripture saith,—"*The good that is done in the earth, Thou, Lord, doest it.*"

CHAPTER 5

Stirpiculture; or,
The Scientific Propagation of the Human Race

Introduced by Michael W. Perry

For it is this class of the unfit, who, on becoming parents, engender in their children the hereditary consequences of their thoughtless marriages, thereby sealing the doom of unborn generations.

Stirpiculture refers to the selective breeding (culture) of special stocks or races of plants and animals (stirps). It is still used that way in gardening, but in the sense that Victoria Woodhull used it (human breeding), it was replaced long ago by *eugenics*. The latter is easier to pronounce, and its root meaning, "good beginnings," made it easier to promote than a word that referred to cattle-like stock breeding. Today, discussions of eugenics almost invariably begin with the later word's derivation.

Woodhull's use of stirpiculture may come from the Onedia Community, a utopian sect founded in 1848 by John Humphrey Noyes. The sect's loose sexual practices made it controversial, and the fact that it delayed and restricted parenthood to 'superior' members created stresses that contributed to its dissolution in 1881. The controversy surrounding it may be why Woodhull knew what she said would create a fuss. Of course, Woodhull also liked taking a martyr's pose. She opened this booklet with a poem by a Scottish writer, Charles Mackay (1814–89), that praised famous people persecuted for their beliefs, as well as champions of Truth and Justice who "toil in penury and grief, unknown if not maligned."

That hardly applied to Woodhull. At the time she published this, she had married (for the third time) into the wealthy British banking family of Martin. The only persecution she was likely to face was muttering from an ill-paid household servant, unhappy to learn that Woodhull held her and those like her "criminally responsible for all the misery from which the human race is suffering through her ignorance of the vital subject of proper generation." (pages 11-12 in the original)

Woodhull opened by comparing society to a building whose architect "corresponds to the Government." (6) Those who suspect she did not believe in democracy are right. She admits that in the first sentence of "Humanitarian Government" (Chapter 6). She hinted here what she later said clearly, that the mass of the population who make up the building's foundation are "rotten and insecure." (5) Later she pointed to a different building, one built "to incarcerate the insane, the idiots, the epileptics, the drunkards, the criminals." (10) If animals, she wrote, displayed "such infirmities and propensities, we should soon exterminate them; and yet we have not thought it needful to take measures to eradicate them from the highest organism, man." That's the essence of eugenics, breeding people for certain traits much as plants and animals are bred.

We can be glad that Woodhull did not demand that the government transform state institutions into killing centers, but her solution is still disturbing. She looks forward to a new "religion of the future, which will be founded on the great truth that the human body is the Temple of God, and will awaken mankind to the awful responsibility of creating His image when unfit to do so."

Her use of religion makes it easy to see why Woodhull's critics were often conservative Christians. Jesus may have loved lepers and cured them, but Woodhull prefers to question whether lepers, as well as the "malformed and misshapen" and those with "brutish faces that crowd the prisoner's dock," should be numbered among "God's creatures." (23) And while Jews and Christians worshiped God, she was more interested in allowing "the piercing rays of science" to "dissipate the mists of prejudice and superstition," so we can create "a race of gods." (23)

This call for a eugenic religion that would sweep the world was a major part of eugenic thinking during its first few decades. You can find examples in *The Pivot of Civilization in Historical Perspective*. Winwood Reade, a mediocre but widely read English novelist, placed his "Faith in the Perfectibility of Man," (Chap. I, 1872). Ellen Key, the famous Swedish feminist, exalted what she called the "holiness of generation" (Chap. X, 1909). In *Herland*, a novel about a eugenic and feminist utopia, Charlotte Perkins Gilman, described how her all-woman society evolved from belief in a "Mother Goddess" to "Maternal Pantheism" (Chap. XXII, 1915). Only with birth control activist Margaret Sanger, the daughter of an avid free thinker, did the religious arguments fade into a vague mysticism about motherhood.

The fact that eugenicists felt a need for a new religion suggests that they knew that no amount of quoting out of context—as Woodhull does here—could turn the Bible into a eugenic textbook. And their scheme was doomed from the start because eugenics is too elitist and coercive for most women.

"Stirpiculture" starts on the next page. "Some Thoughts about America" was part of the original booklet and is included here.

STIRPICULTURE;

OR,

THE SCIENTIFIC PROPAGATION OF THE HUMAN RACE.

BY

VICTORIA CLAFLIN WOODHULL MARTIN,
17, *Hyde Park Gate, S.W.*

FEBRUARY, 1888.
London, England.

ETERNAL JUSTICE.

By Charles Mackay.

I.

The man is thought a knave, or fool
  Or bigot, plotting crime,
Who, for the advancement of his kind,
  Is wiser than his time.
For him the hemlock shall distil;
  For him the axe be bared;
For him the gibbet shall be built;
  For him the stake prepared.
Him shall the scorn and wrath of men
  Pursue with deadly aim;
And malice, envy, spite, and lies
  Shall desecrate his name.
But Truth shall conquer at the last,
  For round and round we run;
And ever the Right comes uppermost,
  And ever is Justice done.

II.

Pace through thy cell, old Socrates,
  Cheerily to and fro;
Trust to the impulse of thy soul,
  And let the poison flow.
They may shatter to earth the lamp of clay
  That holds a light divine,
But they cannot quench the fire of thought
  By any such deadly wine.
They cannot blot thy spoken words
  From the memory of man,
By all the poison ever was brewed
  Since time its course began.
To-day abhorred, to-morrow adored,
  So round and round we run;
And ever the Truth comes uppermost,
  And ever is Justice done.

III.

Plod, Friar Bacon, in thy cave;
    Be wiser than thy peers;
Augment the range of human power,
    And trust to coming years.
They may call thee wizard, and monk accursed,
    And load thee with dispraise;
Thou wert born five hundred years too soon
    For the comfort of thy days;
But not too soon for humankind.
    Time hath reward in store;
And the demons of our sires become
    The saints that we adore.
The blind can see, the slave is lord,
    So round and round we run;
And ever the wrong is proved to be wrong,
    And ever is Justice done!

IV.

Keep, Galileo, to thy thought,
    And nerve thy soul to bear;
They may gloat o'er the senseless words they wring
    From the pangs of thy despair;
They may veil their eyes, but they cannot hide
    The sun's meridian glow;
The heel of a priest may tread thee down,
    And a tyrant work thee woe;
But never a truth has been destroyed;
    They may curse it and call it crime;
Pervert and betray, or slander and slay
    Its teachers for a time;
But the sunshine aye shall light the sky,
    As round and round we run;
And the Truth shall ever come uppermost,
    And Justice shall be done.

V.

And live there now such men as these—
    With thoughts like the great of old!
Many have died in their misery,
    And left their thought untold;
And many live, and are ranked as mad,
    And placed in the cold world's ban,
For sending their bright far-seeing souls
    Three centuries in the van.
They toil in penury and grief,
    Unknown, if not maligned;
Forlorn, forlorn, bearing the scorn
    Of the meanest of mankind!
But yet the world goes round and round,
    And the genial seasons run;
And ever the Truth comes uppermost,
    And ever is Justice done!

STIRPICULTURE;

OR,

THE SCIENTIFIC PROPAGATION OF THE HUMAN RACE.

THE goal to be attained, is the motive which impels people to action. Scarcely any action is performed without an end in view. Therefore mankind has acquired the habit of seeking for the ulterior motive which actuated this or that deed. All civilized nations have a code of laws. The motive which induced the framing of these laws was the necessity of having fixed rules which would regulate the aggregations of families into societies and states; the goal to be attained was the commonweal of the people.

Sociology may be compared to the construction of a building: the myriads of poor are the foundation; the rest of the structure corresponds to the different grades of society. The last could not exist without the first named. It is the struggling masses who are the foundation; and if the foundation be rotten or insecure, the rest of the structure must eventually crumble. We must not consider the durability of the structure from the upper portion, that is, the upper classes, to whom the existing social condition is well enough; but we must look to the foundation of the structure, or, in other words, the millions of wretched humanity who are daily increasing and whose condition is becoming inevitably more miserable. The architect of this building corresponds to the Government; and, before he commences to build, he must be quite sure that the foundations are all right, so as to give strength and support to the whole. To be a great architect is to be highly endowed. The ancient Greeks realized the great importance of this eminent ability, whether it was as the architect of a building, a government, or, above all, as the architect of a human being. The Greeks have reflected upon modern nations all that is highest in classical art. Their then existing popular government brought to perfection philosophy and literature, and was a masterpiece of administrative construction; and history tells us of instances wherein their perception of the beautiful was so keen of what the human body might attain, that they brought the finest statuary and works of art before the pregnant mother to impress her that she might create a perfect model of human architecture.

We shall always find people devoting their lives to perfecting beautiful orchids or some other rare plant; astronomers are searching the sidereal universe to find some new star; amateur breeders in the highest circles of society are devoting their time and their money to

7

perfect their horses and cattle; farmers employ different food for the different effects they wish to have in their sheep,—fine wool, or excessive fat, &c.; all the influences of temperature, air, food, and external surroundings, are brought to bear to perfect the animal and vegetable organisms: but to waste a moment's reflection over the solution of perfecting so miserable a creature as man, what he may become, to what standard he may attain—impossible, the discussion is vulgar. The consequence is that people are not shocked to read and discuss the terrible crimes which may be committed by so low and vulgar a creature as man, or the horrible descriptions of want and misery accumulating on every side among the poor; but they are shocked at anyone's daring to talk of the causes that made this crime and misery possible.

Even the discussion of these social evils can scarcely mitigate the degraded condition of the people whom it is most necessary to reach. Among the poor, to all that appeals to the worst side of man's nature is given full scope. In the terrible fight for existence, they are obliged to work hard all day, and sometimes far into the night, having no time to pause, to consider the terrible evil that they are daily making greater by this crime of reproducing in their offspring their own debilitated condition both of body and of mind. They are perpetuating the hereditary curse from which they are daily praying that death may release them; but from which they are too weak, morally and physically, to abstain from cursing their children with. And these children have not only the hereditary instincts of crime to contend against, but are made familiar from their infancy with vice of every description. And when we read in the daily papers, that crime among children is becoming more prevalent, who is responsible? When our cities are counting their outcasts and paupers by the tens of thousands, to whom are we to look for relief; when those in authority, who represent the people, remain apathetic, or say, when appealed to, that they are powerless to offer a remedy?

In the pulpit, and in every department of life, leaders and teachers are shutting their eyes to this growing evil; but still they say that there has been no suffering like the present; and they make charitable appeals for countless starving, ignorant people, regardless of the causes that make these charities possible. And yet they insist that the people are not prepared for the discussion of the scientific propagation of the race. *The time never was nor ever will be ready for the cowards who fear to be included among the agitators of an unpopular subject.*

It is man who requires to be perfected; and in order to do this we must first understand the laws operating upon his social and physical condition. First, that condition of sociology which made certain human laws compulsory, and which created the necessity for the erection of public buildings to enforce them.

Moses studied for thirty years at the court of Pharaoh, and was initiated by the priests into those laws governing human society which preceded them.

With this knowledge, he constructed laws and institutions which have been more or less modified and adapted by each succeeding government according to the degree of civilization that each nation has attained. The Ten Commandments have become an integral part of the codified laws. The Government does not leave these commandments as optional moral laws, but makes them obligatory. For example, the law says:—

Thou shalt not kill.

Thou shalt not commit adultery.

Thou shalt not steal.

Thou shalt not bear false witness against thy neighbour—that is, thou shalt not perjure thyself. If you break these laws you must suffer the penalty. The Government has the authority to enforce obedience, and therefore immense capabilities of forming the character of its subjects. By the very fact of punishing these offences, it creates a repugnance in the minds of people; and in successive generations, it is not so much the law which is the restraint, as the association that has been engendered with these crimes. It has created a horror in the human *mind* of murder, of adultery, of theft, of perjury. Why cannot it go further and add something towards perfecting the human *body*? Let the Government incorporate as part of its laws the following commandments:—

Thou shalt not marry when malformed or diseased.

Thou shalt not produce His image in ignorance.

Thou shalt not defile His Temple.

Let it make it a crime against the nation, as it is

10

against God's Divine laws, to break these commandments, to thrust the results of their ignorance on the body social to be supported at the expense of the nation. Then in successive generations these crimes, too, may be regarded with horror and repugnance. It is no longer a question of expediency, but one of absolute necessity, which the Government will be obliged to deal with.

What is the commonweal of the public? It is to inaugurate such laws as will react with beneficial effects on its subjects. And as population increases and intellectual growth advances these laws should not remain stationary, so as to retard progress, but should be revised to keep pace with the times. The laws which were made to legislate for our forefathers, should not be allowed to become our legislators. Inertia is not progress. It is owing to the ever advancing nations that civilization has reached its present altitude. The most conservative throughout nature are ever the slowest in development.

We build institutions in order to incarcerate the insane, the idiots, the epileptics, the drunkards, the criminals, &c. If the lower organism of animals were subject to such infirmities and propensities, we should soon exterminate them; and yet we have not thought it needful to take measures to eradicate them from the highest organism, man. All such propensities have been contracted or acquired by the parental organism, or during the life of the individual itself, and have become hereditary in the offspring, reproducing itself

11

so exactly as to develop itself at the same age in the offspring as when acquired by the parent.

What will the future enlightened ages think of the women who are calmly looking on, oblivious of the fact that their husbands are at work erecting the very edifices in which they, in their ignorance as human architects, may see their offspring incarcerated; or if not their own children, some other mother's child? And in after years, in visiting these buildings, they will see the result of their ignorance in the brutish faces of the inmates with the legible stamp of hereditary sensuality and vice. Every building erected to meet the exigencies of this ignorance will be a reproach to all womanhood. Do people realize, that nowhere on the face of the earth is there a building erected to teach people how to perfect the human body? The sooner mankind awaken to the all-important truth, that you cannot force people into moral conduct, into a better social condition—that you must educate them—the sooner we shall have emptied the buildings erected to contain those monstrosities.

The phase of civilization which envelops in mystery the origin of life conceals within itself the greatest incentive to the very crimes that intelligence would dissipate. Well did Sokrates define virtue as knowledge, and vice as ignorance. This truth should be brought home to every woman, and she should be made to feel that she is criminally responsible for all the misery from which the human race is suffering through her ignorance of the vital subject of proper generation. The vulgar prejudices of a narrow-minded people have, in woman, cloaked with ignorance the fundamental principles of life in which our mothers should be most educated—an ignorance which is powerless to inspire in their husbands and sons that chivalry and respect due to every other mother's daughter. The mothers are incapable of teaching their children what they themselves have not been taught, leave their sons and daughters in ignorance of those dangers which await them at the very threshold of the struggle for existence, leave them to take their chance in this unsympathetic, pitiless world, leaving them to weep tears of blood over the dying embers of a misspent life!

Mankind has forgotten the lesson taught by the Hindus five thousand years ago—that to degrade or to oppress woman involves the physical and moral degradation of man; that the assumption of superiority and tyranny of the master, which for ages man has assumed over woman, has almost extinguished that Divine spark in her which alone has the power to regenerate humanity.

We cannot over-estimate the influence for good or evil that the mother has over her child. Who can train with such loving tenderness the shooting tendrils of the young inquiring mind? Who can with such unerring judgment instruct the child in those dangers which surround the developing manhood and womanhood? Who is so capable of awakening the soul to a higher life? And when this is realized, and superstition and ignorance give way to a reign of intelligence, and the full comprehension that there is nothing so vulgar as ignorance, then medical statistics will not have to tell us that one half of our sons on reaching manhood are unfit to become fathers, and that one half of our daughters on reaching maturity are unfit to become mothers. For it is this class of the unfit, who, on becoming parents, engender in their children the hereditary consequences of their thoughtless marriages, thereby sealing the doom of unborn generations. The marriages of such individuals produce epileptics, idiots, neurotics, insane inebriates; and by far the larger number become criminals. Nearly all the crime committed is the result of the inherited weak blood, or a malformation, or disease of the brain. It is familiar to everyone that when the brain is injured a corresponding loss of intellectual activity supervenes, causing in some cases a total lapse of memory, or hallucinations which occasion the most extraordinary actions on the part of its victim. If we know the cause producing this effect, we regard them as irresponsible beings and try to alleviate their sufferings; but where the cause producing this effect is unknown to us, where these hallucinations and extraordinary actions are the result of an hereditary diseased formation of the brain, or injurious conditions of life reacting on the brain, we regard them as hardened criminals and hold them as responsible citizens. It only proves to us how little the psychical as well as the physical part of man is understood. We should not treat these unfortunates as criminals, as responsible beings; we should treat them as having a diseased mental condition. For it is the co-operation of the different parts of the body which insure the organization of the whole; injure one part, and the whole suffers; debauch the stomach, and the whole is debauched; disease the mind, and the whole is diseased; outrage nature, and monstrosities are the result. By the magnitude of these social evils we are overwhelmed, but of their origin we have been neglectful.

When we are able to make the study of man take precedence over all others; when we are able to reduce the psychical part of man to a science; when we are able from these conclusions to perfect his moral and physical condition; when we are able to judge our own actions with the same justice as we judge another's; when we are no longer the slaves of habit; when we realize that within us is a little community which requires to be properly governed, so that anarchy will not take possession, thereby causing destruction to the whole; when we realize that passion is the barbarous state of man, and reason is the civilized state, then we shall no longer look into the vacant faces of people and see no response to the question of decaying nations. Why has this crime of ignorance been allowed during these thousands of years of civilization?

We have only to visit the poorer quarters of our towns and see the slaves of drunkenness and disease, to realize upon how slight a basis is built the security of

15

our modern civilization. What a lesson might be taught in administrative government. For no matter to what perfection we may attain in mechanics, art, or sciences, man is the concentric pivot around which all revolve, and without which everything is an illusion. That immeasurable space, with its innumerable stars and glowing suns, piercing the mists enveloping worlds, and all the elements waging the fierce battle of life, are only realities in as far as man's intellectual power is capable of understanding them as such. The faults existing, and inharmonious notes producing discord, originate in man. The more the brain of man develops, the more the unity of inanimate and animate nature will be understood, so that ultimately the gulf dividing the two will be spanned, and a violation against one of nature's laws will be impossible.

16

"Know ye not that ye are the temple of God?"

"If any man defile the temple of God, him shall God destroy; for the temple of God is holy, which *temple* ye are."—1 *Corinthians iii.* 16 *and* 17.

THE votaries of truth are ever tortured and crucified by the bigoted and ignorant. The philosophers of ancient history realized greater truths than they dared proclaim to an ignorant populace. Preachers, all over the civilized world to-day, are in advance of their congregations; but they teach their philosophy, as of old, according to the calibre of the people, and not the truth that they realized. The spirit may be willing, but the flesh is weak. In the past, divines have been able to wield an immense power through the ignorance of those to whom they preached. To-day people are thinking for themselves, and the more they comprehend, the more the Church loses its power; and it will continue to do so, unless it have something better to offer than hitherto.

The time has come for the strong food of which St. Paul speaks:—"I have fed you with milk, and not with meat: for hitherto ye were not able to bear it, neither yet now are ye able." (1 Corinthians iii. 2.) "For everyone that useth milk is unskilful in the word of righteousness: for he is a *babe*. But strong *meat* belongeth to them that are of full age, even those who by reason of use have their senses exercised to discern both *good* and *evil*." (Hebrews v. 13-14.) Now, St. Paul meant that intellectually we are as yet infants, too weak to grapple with the truth that must come to us before we shall be able to solve the mysterious problem of life. Has habit paralyzed the impulse of nobler thought? Is tradition too great a giant to cope with? Must we crush the hunger in our hearts that is craving for the truth, and die like so many beasts of burden? More and more intense becomes the craving desire to unravel life's tangled web. Whence comes the Divine spark that electrifies the clod into animation; and, having spent its force, whither goes the vital element? From the lowest form of animal life we are beginning to recognize the highest type of intellectual growth. Religion is to us the belief in the unseen; and science is the knowable, which, after all, is only the manifestation of the unseen.

We trust no more to those illuminations of the soul which have inspired each new discovery and elucidated many mysteries of science. Yet the spiritual must have had a part in all the great truths that have burst upon the world, though little understood are the ethical laws which govern mankind. That matter can never be annihilated, is the first principle we learn in

18

science. It may be resolvable into compounds, or even into elements; but that it ceases to be a substance is impossible. Once having lived, we can never know death. Death is a misapplied term: "Both animals and plants have their origin in a particle of nucleated protoplasm which not only dies and is resolvable into mineral and lifeless constituents, but is always dying; and, strange as the paradox may seem, could not live unless it died" (Professor Huxley). "That which thou sowest is not quickened, unless it die." (1 Corinthians xv. 36.)

The laws of evolution are, no doubt, the solution of many problems of science; but the exponents of evolution have not been able to produce life, nor to solve this enigma. Five or ten thousand years have not materially changed man. The philosophy taught by Christna and Buddha, thousands of years ago, was grander in its conception than anything we have at the present time. Do we excel the ancient Greeks in art to-day? And again, our laws are simply a repetition of the old Hindu laws dating back more than three thousand years before the Christian era. And in all these years of evolution we have not been able to better them. Five thousand years ought to improve mankind; but the people are in the same depraved condition now as then. If any change has been made it has been for the worse, which proves to us that there is something radically wrong in the social economy.

Faint glimmerings of the truth were perceived by a few redeemers even in the first epochs of history; but it has had to filter through cycle after cycle until woman could be educated to realize her true position as a creator—the architect of the human race. And that Divine martyr, Jesus, foresaw that ages upon ages must roll away ere the human heart could comprehend the true significance of the curse which was marring the masterpiece of God in the Eden of his body.

The tradition that our first parents sinned has been generally accepted: whether, according to the Vedas, it was the Adema and Heva (meaning in Sanscrit the first man and the first woman) on the island of Ceylon, or the Adam and Eve of the Christian Bible. A tradition of which the world makes a fable, but, rightly understood, conveys the greatest lesson ever taught. In the Hindu version, the man was the tempter, and it was at the prayer of the woman that God pardoned man. The Christian rendition makes the woman the tempter. It matters little which it was, but that they both sinned is evident; and, according to the record, for the sin that they committed, all their decendants were to be cursed until mankind should learn to sin no more.

First we must comprehend what the primordial curse really was from which we are suffering to-day, and endeavour, if possible, to remove that curse. We will take the Christian version of the Bible (Genesis iii. 5, 16): "God doth know that in the day ye eat thereof, then your eyes shall be opened, and ye shall be as gods, knowing *good* and *evil*." .... "Unto the woman he said, I will greatly multiply thy sorrow and thy conception; in sorrow thou shalt bring forth

20

children; and thy desire shall be to thy husband, and he shall rule over thee." This was the curse that fell upon the human race when universal man and universal woman were driven out of Eden. No misconception as to the true meaning of this curse can arise: woman, in whom burneth the lamp of God, was to be degraded into an unholy vessel! Each generation has suffered from this curse, and none more than the present. So general is it to-day, that there is not a place on the globe which does not degrade woman—universal degradation! O God! to what depths has thine Eve fallen!

In studying the lives of all Redeemers, from the time of the birth of Christna of Hindustan, born of the virgin Devanaguy, to the birth of Jesus of Nazareth, born of the Virgin Mary, we learn that, although each Saviour was accredited with Divine conception, they were born of mortal mothers. The mothers, through the long anxious months of gestative growth, moulded and stamped the characters of these Divine martyrs. No matter how mythical the conception, it was an awful realistic fact to the woman who gave each child birth. It was she who went into the very valley and shadow of death to usher in this new life. It was she alone who was the architect of the yet unborn being. And through her awakening to the awful responsibility that she has occupied as a creator, down to the present day, will come the redemption of mankind, for her power is just as potent for good or evil on the future unborn generations as it was in the beginning of time.

21

The illuminated soul of Jesus understood this truth, and all His teachings were for the comprehension of this reality. "Destroy this temple, and in three days I will raise it up." ... "But He spake of the Temple of His body." (John ii. 19, 21.)

Even at His Crucifixion the ignorant mob for whom He suffered the martyrdom of death, mocked Him, "Ah! Thou that destroyest the Temple and buildest it in three days! save Thyself."

Christ realized that the redemption of the race must come through the appreciation of the body; and this truth His *Apostles* preached, whereas preachers of to-day, after two thousand years, are still talking of the soul, not having the moral courage to grapple with this truth. And the ignorant cry out, "Rather let us die in our corruption than discuss the procreative principle of life." Hence nature has one fearful blotch to mar her sublimity—a deformed, diseased image of God, rotting away with its own debauchery; scarcely one of His handiwork comprehending the marvellous truth, "Know ye not that ye are the Temple of God, and that the Spirit of God dwelleth in you?" (1 Corinthians iii. 16.)

Jesus, in His Divine person, gave the world the living example of the triumph of the spiritual over the animal. Jesus tried to deliver man from the curse which was put upon Him in the Eden of his body. His teaching was the dawning of a new era. But to this truth the people, whirled in the maëlstrom of their own passions, paid little heed.

22

Sacerdotalism has so twisted and misrepresented every great thought, that mankind know not what is truth and what is error, and superhuman efforts are needed to lift humanity from the bogs of ignorance. Rather give us the stoic philosophy of Epictetus or of Marcus Antoninus than that religion which panders to ignorance. It is the very weakness and cowardice of the few who lead that is spreading desolation throughout the land. We see people cursed to-day with hereditary diseases, hereditary brutish passions, and with hereditary criminal instincts. What can we expect from the man born with alcoholized brain cells but a drunkard? Can we expect anything else than brutish ungovernable passions from men and women, when we consider that the mother-architect during the period of gestation was subject to unbridled passion and brutal treatment until all her capabilities for moulding the character of her yet unborn child for good were destroyed? Can we have anything but murderous instincts from the unwelcome child whom the mother did everything to kill before giving it life, and who engraved upon the child's plastic brain the desire to murder? And when the gallows demands its victim, which one should be hanged? The cruel irony of fate makes mankind the slaves of congenital instincts and congenital deformities. What availeth it to talk to man of his soul, when body and mind are torn by an hereditary disease until reason totters? The law of compensation is inexorable. For religion, men and women have had the courage to be torn to pieces. For religion they have had the courage to be burnt at the stake. For religion, healthy men and women have doomed themselves to perpetual celibacy. The religion of the future, which will be founded on the great truth that the human body is the Temple of God, will awaken mankind to the awful responsibility of creating His image when unfit to do so.

This Temple has been so defaced and brutalized that one can no longer apply to it the word human. Can it be possible that the masterpiece of God is drugged in opium hells scattered over the so-called civilized earth? Are those God's creatures that fill the hotbeds of infamy which swarm in all large towns? Are those God's creatures that people the houses for lepers? Are all those malformed, misshapen beings His image? And are all those brutish faces that crowd the prisoners' dock a part of Him? Happily, men and women are no longer satisfied with food for the body alone; their souls hunger for the knowledge that will give them the mastery over themselves.

The ignorance which surrounds the awful procreative problem makes mankind abject slaves. How long ere the piercing rays of science shall dissipate the mists of prejudice and superstition, leaving us to see, and inciting us to aspire after, that altitude of perfection which, when attained, will make us indeed a race of gods?

24.

SOME THOUGHTS ABOUT AMERICA.

———•◦•———

"While they promise them liberty, they themselves are the servants of corruption."—2 *Peter ii.* 19.

———

THIS is pre-eminently an age of progress. A restless dissatisfaction with the old order of things pervades every heart. The needs of the people grow daily greater, and they will not submit to being ignored. If there be a cesspool in a street, the neighbours do not hastily cover it up so that it may be hidden from the public view. No; they have the very bottom dredged that their loved ones may not sicken and die from the malaria. But the social and political cesspools may go on gathering in the germs of deadly miasma, while each human soul vies with the other to ignore the fatal effects, until the nation is ready to faint under its own corruption. I say social and political cesspools, because one is the outcome of the other.

The government of a nation indicates more than anything else the character of its people. The laws that are needed and enforced determine the degree of civilization which that nation has attained, and the extent to which the people are capable of governing themselves. The laws demonstrate whether the people are the slaves of their own passions, or whether they feel the true dignity of manhood, and are conscious that they have control over a kingdom that they within themselves can make or mar. We have nowhere on the face of the globe a government which aims at perfecting its people. Upon the slightest provocation nations and individuals alike are ready to resort to brute force which was the government of the lowest type of civilization, and from which we to-day are not educated enough to free ourselves.

The best land if left uncultivated grows the rankest weeds, so the country with the greatest possibilities, if governed unwisely, may commit the most fatal errors.

More than a hundred years have come and gone since that momentous day when our Declaration of Independence was signed. What a promise of future achievement! No shackles of autocratic despotism to weigh down the American people! They were free—free to mould their own future greatness.

Let us see how this promise has been fulfilled. The

century has seen many Presidents in office; men who have had glorious opportunities to show the world that our Government was the highest type of human advancement—that they were self-made men, an example of indomitable courage and perseverance, and that through these qualities they had attained the highest office America had to bestow; that they were worthy of the nation's confidence, and would work for that nation's greatness to the best of their ability—to realize, in fact, an ideal republican government. Have we in any one respect attained this? Has any European country any confidence in our officials? Have we ourselves any reliance on the integrity of our representatives? Is it the best man to whom is intrusted the highest office? Is he put in office by the voice of the people, or by the political wire-pullers who have their own self-aggrandisement in view? We call ours a republican government. Does that mean that it is framed for and by the people, or for and by the select few? We see in every department diplomacy and trickery indifferent to the cry of the great mass of starving individuals. We shall have to give an answer to their demands before many years elapse. But unless this problem be met and grappled with soon, America will have one of the most terrible revolutions that has ever shaken the world. A political, social, and moral earthquake is concentrating its force beneath the very foundations of our so-called free country.

The laws of the United States are constructed to

deal with effects only, and do not take into consideration the causes.

Good care is taken that each State shall have its prisons, lunatic, idiotic, inebriate, foundling, and other asylums; but not one building is erected nor one law enforced that would teach the people how not to contribute to these over-crowded receptacles of human misery. And yet the American Government can boast of the surplus in the United States treasury with which it is at a loss what to do.

Never a consideration or thought for the thousands of human beasts of burden who live from hand to mouth, to whom justice, nature, and God are dead letters.

But they are regaled by the Press with a description of what the President has had for breakfast, luncheon, or dinner, how many calls he has had, how many times he has shaken hands, &c. To what great uses is put the highest office that America can confer for the organizing of peace, prosperity, and good will to all!

Let us review the situation. Paupers, tramps, and professional beggars are largely on the increase. Statistics show that the same name will constantly recur among the diseased and criminal classes, and that pauperism is hereditary. Say that one criminal may be the ancestor of a thousand criminals, and one pauper the ancestor of a thousand paupers. Are these not questions that are seriously connected from an economical and social point of view with our Government? Have any of our Presidents tried to meet this question, or in any way tried to alleviate the anxiety of the people?

The social question has reached an acute stage. We may arrest the disease for a short time, but all the more terrible will be the malady when grappled with. We see thousands and thousands of parasites born every year who have no means of subsistence, who are destined to fasten upon their fellow-creatures, draining the vitality and strength of the nation, and precipitating its downfall.

It is not so much to political power that a country owes its greatness, but to that social fabric upon which all laws are based. All of our politicians are ready to deal with the effects, but not one of them is brave enough to penetrate the substratum of society and deal with the cause. A weak and vacillating leader does more to destroy a nation's greatness than does a really bad man. Our Government should be the example to the world, but above its doors of State are written in letters of blood, "*Mene, Mene, Tekel, Upharsin.*" The people want the leader capable of grappling with this hydra-headed social monster and one who understands the terrible urgency of this question. For they will no longer tolerate in power the men who assert that the masses are too ignorant to comprehend the true cause of these social upheavals. The Government should make it possible for the masses to be educated on these vital subjects, that they may no longer in ignorance thrust upon the body social the myriads of paupers, and those who, to eke out a miserable existence, are forced through incompetency for labour to accept the smallest pittance. They thereby exhaust the demand for the skilled artisan, they lower the standard of labour, but above all they curse humanity by doubling their kind every few years, until all our large towns are over-populated. Education and the proper understanding of the procreative principle of life are the only checks to over-population. And this was the curse, not the blessing, which was put upon woman when she was driven out of Eden: "*Unto the woman he said, I will greatly multiply thy sorrow and thy conception; in sorrow thou shalt bring forth children; and thy desire shall be to thy husband, and he shall rule over thee.*" (Gen. iii. 17.)

Yes, read your Bible again, not with the understanding dwarfed and blinded by the bigotry and darkness of centuries, but with the true comprehension that we to-day are suffering from the very curse that was put upon Eve in the Eden of her body. Go anywhere among the most miserable of any community, and you will find there the largest number of improvident marriages. Mere girls and boys, married and not married, become parents before complete maturity. And herein lies the germ of the world's discontent; and the disease is not in the crust which covers the ulcer, but in the core of the ulcer itself. Until that is probed with a daring and skilful hand, socialism and poverty will go on increasing, for the people who will not, or do not know how to, work, are the superfluous, or, in other words, ought never to have been born. For as Jesus turned and said to the women who were following Him, "*Weep not for me, but weep for yourselves, and for your children. For, behold, the days are coming, in which they shall say, Blessed are the barren, and the wombs that never bare, and the paps that never gave suck.*" (St. Luke xxiii. 28-29.) To those in power the little bacillus of socialism has no awful import; it is only when the consequences begin to fret them that they will turn their attention to the germ.

We have agricultural fairs all over the civilized world; each one is competing with the other to breed the finest horses and cattle, and prizes are awarded for so doing. In the animal kingdom in a few years we shall have none but domestic animals, and those in a higher state of perfection. All wandering herds will have been exterminated. But in the human kingdom, vagabondage is on the increase; and even the children of these vagabonds, when taken to reformatories or put into schools, sooner or later find their way back to swell the ranks of that rapidly increasing pauper class, who are lower than animals, having no instinct of domesticity. If the animal kingdom were subject to the debauchery, the foul air, the unwholesome food, the filthy abodes, the prevalent diseases that the human race has to contend with, in five years it would be extinct.

A man could not put on again the clothes of his childhood, he has outgrown them. The laws which clothed our constitution over a hundred years ago, when the American nation was in its infancy, are too small and too narrow in their limits for the intellectual demands of her people at the present day. The people have grown beyond them. To-day is as pregnant with revolution for independence, and as laden with as mighty import as was that memorable day in Philadelphia, when Tom Paine rose to the situation and said : " What this country wants is independence, and I mean revolution;" what the people want now is independence of thought and action and a revolution of old systems and ideas.

The Constitution of America, tattered and torn, has been hallowed by the blood of her noblest sons fighting for freedom. And that same spirit of freedom is only slumbering in the breasts of her sons and daughters to-day. And when she realizes that she is the leading nation of the world, she will rise to the occasion, and shake off the shackles of the Old World's diseased and worn-out social systems which are gradually creeping in and destroying her young life.

VICTORIA WOODHULL MARTIN.

February, 1888.

Chapter 6
Humanitarian Government
Introduced by Michael W. Perry

> *A humanitarian government would stigmatize the marriages of the unfit as crimes; it would legislate to prevent the birth of the criminal rather than legislate to punish him after he is born.*

In *Stirpiculture* (Chapter 5), Victoria Woodhull explained her basic goal, dramatically reducing social ills, particularly crime, by restricting who would be allowed to have children. In *Humanitarian Government* she described what a "humanitarian Government" would do to further that goal.

Keep in mind that when she published this booklet (1890), governments had few means to prevent someone from having children. Surgical sterilization would not come into use until the mid-1890s and would not become common until a decade or two into the next century, when the more 'progressive' states began to pass eugenic sterilization laws. In that era, birth control techniques used barrier methods that could not be coerced outside an institutional setting. Abortion was generally illegal and of little use, since unwilling mothers could easily conceal their pregnancy until the baby was too far along to be aborted.

Efforts were made to ban sexual relations. In 1895 Connecticut passed a law that called for a three-year prison sentence for "Every man who shall carnally know any female under the age of forty-five years who is epileptic, imbecile, feeble-minded, or a pauper." And in 1897 there was an attempt to ban, across the nation, marriage to someone who was feeble-minded or insane. But even the supporters of this legislation realized it would do little. Banning marriage only led to sex outside marriage. Banning sex didn't prevent it.

The only technique that worked was confining people in state institutions where contact with the opposite sex was severely restricted, and that was expensive. Those were the problems Woodhull was wrestling with just beneath the surface of all the booklets reproduced in this book. Sterilization, better techniques of birth control (including some that could be coerced), and legalized abortion all come later, and each would be adopted in turn.

You may want to read the first two pages with particular care and note especially Woodhull's opening remarks. Although she could adopt democratic language when it suited her, she clearly does not believe in "a democratic form of government," and prefers one she calls "humanitarian." (page 3) The reason isn't hard to understand. Not trusting ordinary people as parents, she has no reason to trust them as voters.

Do not let a soft and warm word like "humanitarian" confuse you. Always translate a writer's idealistic abstractions into the concrete actions they require. What she wrote about here are experts deciding how the rest of humanity will live their lives. The fact that a *few* people aren't managing their lives well becomes for her an argument that *most* people need an all-encompassing government that allows a scientific elite to manage the lives of almost everyone. That's her humanitarian government at its rawest. You can think of it as a highly authoritarian late nineteenth-century factory if you like, a factory whose chief product is babies. Aldous Huxley would describe such a place in his 1932 *Brave New World*—a world where babies were grown in bottles and no one, outside a few primitive reservations, is a parent.

You will find the rationale Woodhull's humanitarian government will use to justify its power and actions described at the bottom of page 5.

> ... People as a whole are the reflex of their religious, social, and political institutions. These are the three powerful forces which rule the individuals and consequently mold their characters. These are the sculptors of human souls. (5)

Notice the "people as a whole." Woodhull believed the great majority of people act on "reflex"—meaning instinct rather than reason—and that their personalities are dominated by three forces—religion, society and politics—rather than thoughtful choice. In short, they're puppets on a string. Because these people are not making real choices and shaping themselves by their thoughts, there's no one 'there' to be violated when one set of forces is replaced by another, when her humanitarian government takes the place of whatever religion, society and politics have done in the past. A puppet has no right to complain when the puppeteer changes, particularly when the change is for the better.

Of course "as a whole" also meant there are a select few who are not the product of these forces—people such as Woodhull, who see themselves defying convention and boldly calling for radical change. They believe they (and they alone) reason based on science and facts rather than superstition and tradition. Not being puppets, they feel they have the right to take charge of the puppets' strings, deciding who will have children and how those children will be raised and educated. (Those who are drawn to these sorts of ideas are often the ones least qualified to exercise that power.)

Woodhull's initial remarks about race need to be read in light of her other remarks. In Chapter 7 we point out that she was a racist when it came to the great majority of humanity. Here we see that she believed in what might be called a trans-racial elitism or, as she put it: "The most highly cultivated of any race are my companions. Debauchery, vice and ignorance are my enemies, irrespective of nationality" (3). Notice the clever wording. If the "highly cultivated" were her friends, she ought to be saying that the debauched and ignorant were her enemies. That hostility came out clearly when, referring the Moses and the book of Genesis, she wrote of "the misshapen beings who were numerous even in his time," and asked if we are "to place all these diseased, malformed beings in the same category with those who, in the allegory of the beginning, were the most perfect images of God?" (23) Her elitism isn't that removed from racism.

Elsewhere Woodhull was more specific about the nature of her enemies. The poor man, she believed, does not know good from evil. He is only restrained from beating his wife by the threat of jail.

> Go into the lowest quarters of any city and ask the inhabitants, What is sin? What is duty? What is morality? ... They know nothing about the laws governing their own beings, much less the consequence of violating these laws. But there is something they do know, and that is that if they break a man-made law they will be punished. This wholesome dread keeps them from giving unlimited vent to their animal instincts. (26, see also 49)

On the next page, Woodhull was clear that she not only saw most people as little more than beasts driven by "their animal instincts," she wanted the government to treat them as such.

> ... Punishment is to teach an animal when it has done wrong; it is thus we teach our domestic animals. We train a horse, we whip him when he has done wrong, we pat and encourage him when he does right. In the same way the Government trains us; it is like the bit in the horse's mouth. (27–28)

Woodhull's low view of the poor led directly to a willingness to do almost anything to make them conform. She intended to deal with a ill-behaved "patient" through criminal courts "presided over by a council of scientists" who will rule on who is "curable or incurable," with the latter "confined the same as insane and idiots are now." (31) Shoplift once, and you may get by. Get caught shoplifting several times and a social scientist specializing in shoplifters might rule you incurable and lock you up for life. Do you think ordinary people would refuse to put up with that injustice? You forget that Woodhull dismissed rule by ordinary people—democracy—in her first sentence. This is rule by the Few Who Know Best.

Her idea has caught on. Sixty years after Woodhull wrote of a humanitarian government, the Oxford writer, C. S. Lewis would write "The Humanitarian Theory of Punishment," pointing out the dangers contained within.

> The Humanitarian theory, then, removes sentences from the hands of jurists whom the public conscience is entitled to criticize and places them in the hands of technical experts whose special sciences do not even employ such categories as rights or justice.... Let us rather remember that the 'cure' of criminals is to be compulsory; and let us watch how the theory actually works out in the mind of the Humanitarian.... The first result of the Humanitarian theory is, therefore, to substitute for a definite sentence (reflecting to some extent the community's moral judgment on the degree

of ill-desert involved) an indefinite sentence terminable only by the word of those experts... [1]

Woodhull was no fool. She knew that locking people away was expensive, so she has other, less costly techniques in mind. Remember that she believed most people were the products of three forces—religion, politics and society. She's dealt with religion by creating a new one to replace Christianity. She's dealt with politics by changing the laws and the court system to make it more expert-centered and scientific. Now she turns to the standards that a society sets, standards enforced by stigmatization as well as laws. "A humanitarian government," she wrote, "would stigmatize the marriages of the unfit as crimes; it would legislate to prevent the birth of a criminal rather that legislate to punish him after he is born." (49)

Remember too that making something legal can be as effective a tool of social control as making something else illegal, and it is often far less costly. Legalizing sterilization (*Buck v. Bell*, 1927), which Woodhull praised, or making abortion as easy as possible for the poor (*Roe v. Wade*, 1973) are both considered ways to get rid of the 'unfit' and prenatally reduce crime rates. Woodhull was far ahead of her time in the worst imaginable way.

There is, of course, much more that this humanitarian government would be doing. The parents of criminals, for instance, are to be punished (50, 54) for the deeds of their children. It is an odd demand, particularly when you realize that, if a bank robber is not to be punished because his hereditary and environment made him that way, why should his parents not get off with the same excuse. After all, it is harder to see the 'wrong' in parenting a child who, twenty years later, turns out badly than to see the wrong in bank robbery itself.

But if you think that way, you have missed the entire point of Woodhull's humanitarian government. Crime isn't a personal failing, it is a management problem. It exists because experts have not been given enough power to get rid of it by eliminating the wrong sorts of people, much as a animal breeder weeds out troublesome traits. From Woodhull's perspective, we're still too Judeo-Christian, too moral and too democratic. We need to become eugenic-minded, scientific, and indifferent to the rights of some individuals. There's also a nasty logic behind her idea of punishing parents for the crimes of their child. It provides a rationale for punishing parents *in advance* for a child's crime by keeping them from having that child in the first place.

Woodhull also confused physical infirmity with moral failure. Marriage licenses were to be denied to those who are "malformed or having a transmissible or communicable disease," as if crimes and social ills were the exclusive domain of the ugly or the sick. (55f.) The list of needed reforms she called for goes on and on. It includes hiring "female philosophers to teach mothers" something she assumes most women don't understand, "the full responsibility of mater-

---

1. C. S. Lewis, *God in the Dock* (Grand Rapids: Eerdmans, 1970), 289–90.

nity." We should not shy away from what she believed. This pioneering feminist thought most women were stupid.

I close, as I began, pointing out that Woodhull's agenda had to be broad and intrusive because at that time the means for prevented unwanted births were so primitive. Almost eighty years later, one of those who picked up the torch she lit would have more tools at his disposal and thus more ways to coerce and manipulate. In a medical news magazine, Dr. Alan Guttmacher, a former vice-president of the American Eugenics Association (no surprise there) and at that time the president of Planned Parenthood (notice the link), would warn that if "voluntary means" did not limit population growth:

> Each country would have to decide its own form of coercion and determine when and how it would be employed. At present the available means are compulsory sterilization and compulsory abortion. Perhaps some day a way of enforcing compulsory birth control will be feasible.[2]

The last of the three technologies he mentioned acquired the ability of being "compulsory" with the development of injectable contraceptives such as Depo Provera. Not accidentally, the first series of injections proved difficult to remove when women changed their mind.

Had she lived another half century, Victoria Woodhull would have been delighted by Guttmacher's remarks. Open eugenics of the sort she taught had been discredited by the zeal Nazi Germany displayed for providing birth control, sterilization and abortion to those it disliked (particularly Jews and Slavs), but a crypto-eugenic agenda could be advanced almost as easily by basing arguments on quantity rather than quality—by quoting Malthus rather than Darwin. Unregulated parenthood could be referred to as a "population explosion" and attacked as the primary cause of war and famine rather than as the source of crime and social ills. The result of was the same, a humanitarian government in which, in Guttmacher's words, "each government would have to decide its own form of coercion." But the ultimate goal Woodhull described at the end of her booklet remained the same. To perfect humanity, people were to have their reproduction managed like that of domesticated animals.

> The specific aim of humanitarian Government would be to concentrate all that is noblest and divine in human nature, to strengthen and revivify the life-giving principle (protean matter) into greater and more perfect types of human beings. (68)

---

[2]. "Outlook," *Medical World News* (June 6, 1969), 11.

HUMANITARIAN GOVERNMENT.

BY

VICTORIA C. WOODHULL MARTIN.

LONDON:
1890.

HUMANITARIAN GOVERNMENT.

I DO not believe in a monarchical, nor in an aristocratic, nor in a democratic form of Government. I believe in a humanitarian form.

The aim of a humanitarian Government would be to promote the physical, consequently psychical, well-being of its subjects. It would be applied scientific knowledge for the benefit of humanity. It would have a standing army, but it would be to wage war against debauchery and crime. It would have a national flag, but that flag would wave for the perfecting of human beings, making no distinction of race under its banner.

The day has passed for race distinction. All advanced nations are rapidly becoming cosmopolitan. It is now the most generous and noble who are my brothers and sisters. The most highly cultured of any race are my companions. Debauchery, vice, and ignorance are my enemies, irrespective of nationality.

4

The aim of humanitarian Government would be to organize and unify all that is noblest and purest of all nations against debasing conditions. It would inaugurate an age of Reason by appealing to the populace with noble and humane examples. When a Government is truly humanitarian it will inspire confidence, trust, love, sympathy, which are the fusing elements that make a people combine to give power to the executive. A humanitarian Government would not recognize any caste, except that of personal worth.

The laws which may be beneficial for two or three generations may be totally inadequate for the fourth. I recognize the law of evolution in me and that no single thing remains stationary. I cannot repeat exactly the experience of yesterday, were the same circumstance to repeat itself. I have changed ; I am a day older. Molecular changes have taken place. The identical circumstance in all its details will never be repeated.

A humanitarian Government, whilst recognizing that a Government must be sufficiently stable to insure security, would be essentially a progressive policy. Every new law is on trial, and if found inexpedient or nugatory in its effects should be

5

repealed. Better change a bad law than to have its evil effects vitiate society by not so doing.

"We," to quote Aristotle, "form the *character* of our citizens by enforcing habitual practice"—for good or bad. We want the actual practice of justice not the theory. We want the realization of liberty in the actual lives of individuals when they become masters over, instead of slaves to, their passions. "Who is free? The man who masters self." (*Epictetus.*)

If the present is the result of the past, the future will be the result of the present. Legislators must awaken to this fact in formulating laws. No light task, no thoughtless duty is herein entrusted, but the welfare of humanity.

The same as the prevailing ideas of an age indicate the sentiments, needs, and religion, so the laws that are needed and enforced intimate the degree of civilization of that age. People as a whole are the reflex of their religious, social, and political institutions. These are the three powerful forces which rule individuals and consequently mould their characters. These are the sculptors of human souls!

6

Religion is now passing through the period of analysis. We laugh at the conceits, the delusions, the self-imposed tortures of religious fanatics. We look upon religious ecstacies as a species of dementia. Idols, images, ceremonies, are so many tricks to play upon the credulity and subjugate the will of the ignorant. Many say, If it does any good let it be, we must have something to hold the ignorant masses in check.

If we ask what is religion, the answer would be a belief in God, who is supernatural, who is supreme, who is all-powerful. Above all, one who will *punish* if you disobey His commandments. The last gives us the clue of the origin of religion, to inspire awe, to fear.

The instincts of prehistoric man are not yet burnt out of our organisms; these are impatience of control and revolt against anything which restrains. Any power which undertook to do this has had to inspire terror enough to control or to check the evil and to develop the good. Evil is that which is injurious to the species, good that which is of advantage to them.

The spirit of religious commandments had been taught by the philosophy of experience. The religious laws had a reason to be, they were not the

7

figments of the imagination representing the hallucinations of the superstitious, they were inevitable conclusions which had been arrived at by attempting to regulate roving tribes from the family into societies for mutual benefit.

The principle was to conquer and subjugate by inspiring fear rather than by shedding blood.

The supreme power assumed different phases; fire, sun, superhuman beings each in turn was personified as a god. He figured in so-called myths depicting the passions to make them human. These myths were representing the struggle between the animal man in some form and the human man.

Laws, whether religious or political, are made to rule human beings and animal beings; we have to study, therefore, what made animal man human, and when he ceases to be human and reverts back again to animal man. Legislation has not only been for human beings but for animal beings with only their animal instincts developed—nutrition, propagation, and self-preservation.

In measure as human man or animal man acquired the supremacy, in measure each nation became great or sank into decay.

8

About ten thousand years ago people had their five senses. They could see the earth beneath them, the life around them, and the sky above them. They did not look at these things mechanically as is shown in the four Vedas—Ritch Veda, Sama Veda, Yadjou Veda, and Atharva Veda.

There has been no new sense given to humanity within this short space of time, only aids to our senses, in new inventions, microscope, spectroscope, and mechanical improvements for scientific investigation.

Why should we conceive that the people who lived ten thousand years ago were fools? Our senses are just as liable to deceive us to-day.

Suppose another nation were to get their idea of our modern art literature, scientific discoveries, morality, and of God, from vagrants, peasants, and emigrants, I think their notions of existing phenomena would be as crude and barbarous as any of those of antiquity. When a seed of truth is sown, it takes sometimes years, and even cycles before it reaches fruition.

We cannot ignore one germ of truth. In all these ages, a great many truths have been born unto the world: but some fell among the ignorant, and with all the fury begotten of ignorance, were distorted and rendered powerless, others fell among the willing but weak, and have only been the property of the select few, while others fell among the courageous unto death, and have blessed humanity.

Those philosophers who compiled the Vedas must have had a great insight into human nature in their time. All philosophers have been students of human nature, so we must not throw this book or that aside as worthless because there may be that germ of truth in it, which, if planted in good soil, may bless humanity.

If salvation was taught, there was something to be saved from.

Salvation is the victory over self. This cannot come without knowledge. People are saved in part whenever they receive a new truth and live it in their lives, and are saved completely when their soul is awakened to the whole truth.

Scientists seek the cause why we are vertebrates, why we have organs similar to other mammals. May we not seek the reason why this myth, or that, developed the human in us?

The very first battle was between God, the creative principle of life, and the devil, the desecrator of this principle.

10

We have the myth where the animal man or demon of superior brute force was carrying off or overcoming the female, and the human man always came to her rescue and saved her.

The animal man was depicted as half man, half beast in ancient India, and we to-day can see in the museum at Olympia, chiselled by Greek sculptors, "A Centaur about to carry off a woman, whom he holds with his left hand and right forefoot, while she in her struggles seizes him by the hair and beard. With his right hand the Centaur defends himself against Perithoos (human man) who advances to the rescue with his battle-axe raised." The best preserved group is the "Woman who has sunk on her knees, while the rearing Centaur clutches her hair with his left hand, and holds her fast with a hoof on her breast. The HUMAN part of the Centaur is wanting."

Thus in Greece, thousands of years later, we see the same forces at work to celebrate the victory of the human man.

The priests of ancient India realized how necessary it was to strike terror into the hearts of men-beasts to protect the females. It became the theme of poems, Ramayana, and many others. It even went

11

to the extreme, that the saviours or gods were to be conceived by the Holy Ghost.

Have we ever comprehended what is meant by the Holy Ghost, and why Jesus was so persistent in insisting upon the fact that a sin against the Holy Ghost was unpardonable ? With such a sword of Damocles hanging over our heads, we ought surely to comprehend what is meant by the Holy Ghost, the third part of the triune, to which there can be no remission of the penalty: " The mystery of Godliness."

In the Vedas, Siva, or the Holy Ghost, was the spirit of fecundity or generation, or principle of life, decomposition, and death. And undoubtedly this is the derivation of the term "Holy Ghost." This idea permeated primeval religious sects from the earliest ages. We read that Christna of Hindustan was conceived by the Holy Ghost or Divine essence, that Jesus of Nazareth was conceived by the Holy Ghost. And, without doubt, the people who proclaimed this fact to the world comprehended that it meant the life-giving principle.

Hindoo tradition speaks of the Holy Ghost which moved on the face of the waters at creation and imparted life.

12

Moses had evidently studied heathen mythology before he gave us the description of the act of creation by the Divine Breath, which is the Holy Ghost, into Adam, by which he became a living man.

We must acknowledge that all these give us a clear idea of what was understood in the beginning by the Holy Ghost. And we commence to understand why a sin against the Holy Ghost was unpardonable. We begin to comprehend why there are so many diseased, deformed images of God standing as an open accusation against the ignorance and superstition which from the beginning has enveloped the procreative principle of life. And this is the stumbling-block over which humanity is wrecked.

Every human development we see deified: "Pallas the goddess of Wisdom," "Athene goddess of Chastity," Hercules hero of battles. Every sentiment was appealed to, instincts intensified, to awaken into life powers of comparison between right and wrong.

In the Vedas, we find that artificial selection was carried on to enhance or intensify these propensities thousands of years ago.

The sages who made these laws were deep philosophers; they knew, if the priest married a

13

woman morally and physically healthy, she would produce children who would do honour to Brahminical institutions. Did these great philosophers realize that one was the necessary sequence of the other?

We see scientific propagation practised in India, when it rose to the height of its glory, and was the wonder of antiquity, the reign of the wholly human man; and we see it in Greece, where it developed one of the finest races the world has ever seen.

I quote from Jacolliot's "La Bible dans l'Inde," an extract from the Veda: "Let the Brahmin marry a young Brahmin virgin without spot. Let him not seek a girl of *evil manners* or *unhealthy*. The wife whom he shall choose should be *well made*. Let him shun women of *impure and vulgar race;* their contact shall *defile him*, and thus shall be the *cause* of *degradation* of his *family*. The woman whose words and thoughts and person are pure is a celestial balm. Happy shall he be whose choice is approved by all the good." (Manou, lib. iii.) "It is ordained that a devotee shall choose a wife from his own class. Let him take a *well-formed* virgin, her hair fine, her teeth small, and her limbs charmingly graceful. Let him shun those who neglect the sacraments, who do not produce male children, or whose *parents* are *afflicted* with *defiling maladies*."

What stock-breeder of the present day could add much to the preceding?

The first principle insisted upon was to keep the breed pure. What law-giver have we to-day who equals this in justice, mercy, charity to the yet unborn subject?

A great number of the creeds of ancient India took root in Greece, where the gods of antiquity became the inspiration of a new race.

Scientific propagation was carried on under a new guise by the most astute law-giver the world has ever seen—Lycurgus. I used to ask myself how was it possible to make such laws as he did, and above all, to have them observed.

Scientific propagation is no easy task even in our day. How was it accomplished?

I quote the following from Pausanias: "And it was when Agesilaus was king that Lycurgus legislated for the Lacedæmonians, and some say he derived his laws from Crete, others that he was instructed by the oracle at Delphi. And the Cretans say that their laws came from Minos (Manou, the law-giver of India; Manes, the law-giver of Egypt; Moses, the

15

law-giver who led the children of Israel), who received *divine* assistance in codifying them.

"And it seems to me that Homer has hinted as much in the following lines about the legislation of Minos: There too was Gnossus, the great city where Minos reigned nine years, the bosom-friend of great Zeus."

I am more inclined to believe that Lycurgus derived his laws from Crete.

The most ancient records given of the Olympian games are that they were first founded by Zeus, who came from the island of Crete, and were afterwards reorganized by Lycurgus and Iphitos. Even when these games were first founded, it was called the Golden Age. And as the Cretans say their laws came from Minos who received them by *divine* inspiration, and the intimate friend of Zeus, which meant God, what more natural than when these games were reorganized by Lycurgus and Iphitos they should receive *divine* assistance; especially as the country was being ruined by civil wars.

This myth is a few thousand years later, but it has all the family traits of its ancestors. It was to direct that pent-up energy, which must find vent, into a channel for good instead of bad. When the Olympian

16

games were made a national affair, by giving an impetus to all that was manly and healthful, the very mainspring of a nation's greatness was set in motion. The ideal became perfect health, strength, vigour, courage, and every inducement was given to promote physical development — success was followed by pleasure of the gods, praise of their fellow-creatures, honours reflected on their families.

Their states and their countrymen used to testify their gratitude by triumphal receptions, banquets at the public expense, and sometimes exemption from taxation. Statues were erected, which in case of triple victory were allowed to bear the features of the victors. Champions dwelt at Olympia at the public expense. And not only athletes were represented, but artists, sculptors, orators, heroes, were given ovations.

It no longer surprises me that that age produced so many great philosophers. Socrates, Plato, Aristotle, produced a Platonic Republic. These philosophers had the living examples before them of what can be done when energy is expended in the right direction.

It was the Revolution of '93 that made possible Comte's Religion of Humanity. Books had been written on the rights of the people. Libraries encumbered with grand theoretical works. Dust-eaten, mouldy, there they remained upon the shelves! But when the blood of '93 washed the streets of Paris, it spoke in tones of thunder of the rights of the people, of the Religion of Humanity.

Religion has always represented abstract laws of which the Government is the concrete application.

Religion, if carefully analyzed, is merely an aggregate of ideas which evolve a doctrine. As long as the ideas or principles are consistent they have the power to control our actions; but from the time reason and religion commence to battle, from that moment religion loses the power to restrain.

The ideas of Confucius do not pretend to be a religion, and yet his followers, numbering over 100,000,000, have found it all-sufficient. He taught, "The knowledge of one's self is the basis of all real advances in morals and manners." Animal man must have developed a great way towards human even then, to comprehend what was meant by morals and manners.

Each religion has been the outcome of some social necessity. All martyrs who have had any influence in formulating a religion have been keen observers

18

of human nature. These philosophers have seen the greatest need, and from that have formulated a religious law.

Where anarchy and lawlessness reigned, they said: "Do not to your neighbour what you would take ill from him" (*Pittacus*, 650 B.C.). "Do unto another what you would have him do unto you, and do not to another what you would not have him do unto you. Thou needst this law alone, it is the foundation of all the rest" (*Confucius*, 551 B.C.). "Return not evil for evil" (*Socrates*, 469 B.C.). "Pardon the offences of others, but never your own;" "The noble spirit cures injustice by forgiving it;" "It is a kingly spirit to return good deeds for evil ones;" "Be at war with men's vices but at peace with their persons" (*Publius Syrus*).

Where the poor were oppressed and suffering: "It is easier for a camel to go through the eye of a needle than a rich man to enter the kingdom of God."

Religious rites with regard to ablutions we find in all ancient religions, to enjoin the necessity of cleanliness.

Where there was licentiousness and debauchery, to be carnally minded is death which was taught by

Christna of Hindustan and Jesus. "Blessed are the pure in heart for they shall see God" (*Jesus*).

Where people were the slaves of their passions, to control our thoughts was taught by Christna, Jesus, and Marcus Antoninus. "The happiness of a man's life depends upon the character of his thoughts" (*Marcus Antoninus*, Stoic Philosopher).

"Cast thy bread upon the waters and it will return after many days"—this precept was not so much impressed with any idea that they would receive the equivalent of the service rendered, but by doing good deeds one's own nature becomes enriched and benefited, and ultimately a higher meaning to the word human is evolved.

Every philosopher realized the great importance of preaching against egoism: "None of us liveth to himself, and no man dieth to himself" (*St. Paul*). "Egoism of all human passions is the most difficult to overcome."

"Judge not lest ye be judged," instilled forbearance.

"Love ye one another," benevolence.

The keenest and divinest insight into human nature was possessed by Jesus. It was that great heart which could feel and sympathize with the whole

20

human family, it was his clear insight which originated one of the grandest humanitarian laws ever formulated. On looking into the faces of those around he saw the sin they would condemn in another indelibly stamped on their own countenances: "Let him without sin cast the first stone."

Every new discovery and application of a moral law, every new discovery of some far-reaching scientific law, have all their influence in developing the human man.

Facts exist prior to their identification and verification by scientists, but by being brought together we derive general laws which lead on to higher forms of truth, thus add their quota towards building up the intellect of humanity.

Religion does not remain inert any more than any other product of nature. It ought to be the concentration of nature's laws, which will strengthen all that is best and purest in human character. And as knowledge increases, and the laws of causation are gradually being unfolded to the human mind, some new duty is enjoined, some new obligation is imposed, and some more definite ideas of right and wrong are evolved.

21

The Christian teachings of Jesus were an improvement on the old Mosaic laws of Moses. The conditions of the times were different.

The principle of each religious law was to awaken diviner impulses; they were practical in meeting the needs of the time. Their aim was to humanize the passions.

Altogether the influence of religion has been great in civilizing mankind; because the more the sympathies are developed, the more the innate qualities which distinguish man from the savage is unfolded. That is the God element.

Religious laws, based on the practical needs of humanity, have done good; but there have been many misconceptions derived by professed teachers from religion which have done much to retard progress.

The teaching and impressing upon people that they can do whatever they like, if they only repent they will be saved, has been an incentive to yield to animal passions and appetites, in the belief that if they reform at the end they will be received into the kingdom of God pardoned and purged of all impurities. Never was there a greater mistake, and never was there a greater stumbling-block in the path of progress.

22

If there be a hereafter, the spiritual part of man will find its level in eternity, as it finds its level here. It is a law of nature that a stream never rises higher than its source.

Modern Christianity has devoted all its time and labour in saving souls. I devoted all my time and labour in saving bodies and teaching; if salvation come at all, it must be through the just appreciation of that body—that to defile His temple involves spiritual as well as physical death.

Religion condemns a man as directly responsible for sin by the deliberate act of free will. I taught that sin is simply a sequence of heredity and education manifesting itself through the diseased organism of the perpetrator—that the larger part of crime is the feculence of a distorted social condition by ignoring scientifically demonstrated natural laws.

"Let us make man in our image, after our likeness." Have we ever paused to think over these words? What could be more inspiring than a human being perfect, physically, morally, and mentally? and what an outrage on nature when medical statistics say that there is scarcely a healthy, sound man or woman, and when we consider how many deformed beings are born every year!

23

Moses must have been at fault when he wrote Genesis, because he did not take into consideration the misshapen beings who were numerous even in his time. Are we to place all these diseased, malformed beings in the same category with those who, in the allegory of the beginning, were the most perfect images of God? Or shall we be compelled to class them as the sin against God when driven out of Eden?

Moses understood half the truth, but he could not give it to the horde of slaves and pariahs, made so by generations of inherited effects of servitude and oppression, any more than preachers dare speak the truth to-day to the ignorant. Moses could only give the best he dared in allegory—the two cherubim, the eyes; the flaming sword, the tongue; the serpent, lust.

How many centuries before the descendants of these people could develop into freedom of soul and body? If the people could have been taught in plain language, instead of in allegory, the truth regarding the sin against the creative principle of life would have been born unto the world ages ago.

"Blessed are the pure in heart." "Act in such a way that your conduct might be a law to all

24

beings." Thou shalt not marry when malformed or diseased. Thou shalt not create His image in ignorance. Thou shalt not defile His temple. Do not think evil lest you become evil. These should form part of every religion, be it old or new.

In the Marriage Service of the future the question will not be asked who giveth this woman away? or does anyone know any just cause why, &c. The question will be, how have *you* guarded the vital law you ask me to seal? What use have you made of it? before I give you the permission to become creators, certainly you would not ask me, a minister of the Gospel, to call down the blessing of the Church on a prospective *crime*.

The sooner we awaken to the fact that the more we bear in mind nature's laws in formulating artificial laws to rule human beings, the more certain we can be that they will be enduring and will meet the requirements of mankind.

There is a strongly developed religious sentiment in human beings, the aim is to purify this belief of the dross and direct it into a channel for the higher purposes of humanized life.

It should be the aim of anyone who has any influence in disseminating religion to assemble

25

together natural laws which will regulate our daily life, and will have the greatest good mentally, morally, and physically.

The new religion which will be based on scientific truths will idealize the good, the pure, the great in the individual. Our perception of God is exalted according to the greatness within us. It requires an elevated mind to perceive the beauty and harmony in nature; so to appreciate or to understand Godliness we must have these elements within ourselves.

26

Law is an educational process. It has the power to stamp an action as a crime. Law is applied morality, and the principal means of its diffusion. Moral force is not sufficiently developed in people to deter them from doing an unjust act. Go into the lowest quarters in any city, and ask the inhabitants, What is sin? What is duty? What is morality? Of ethics, they know nothing. They know nothing about a God who leaves them in such a state of misery. They know nothing about the laws governing their own beings, much less the consequences of violating these laws. But there is something they do know, and that is that if they break a man-made law they will be punished. This wholesome dread keeps them from giving unlimited vent to their animal instincts.

This is the power which educates them in morality. This is the power capable of civilizing them when nothing else can influence. For thousands of years this agency has been at work moulding the character of human beings. To the uncivilized man the larger part of what we call crime is no crime; he has not been punished for committing these offences.

The effect of punishment varies with the individual.

27

There is as much difference between human beings in this respect as there is between different species of animals. One form of punishment may have no effect on one, whereas with another the effect may be terrible. Judges have experienced this frequently. A prisoner may be sent again and again to prison to which he has grown indifferent, but if ordered a number of stripes with the lash the effect may be permanent. It only enters into consciousness and has an effect by the physical pain of their bodies with this class; the higher nervous centres are incapable of receiving any other impressions. With another class loss of liberty, disgrace, loss of caste would have terrible effect. Some may feel intensely by even the thought that their honour or their word could be doubted; some are so highly organized, so sensitive, that coarse or vulgar surroundings, or harsh words, or unjust suspicions may almost kill them. Remark the different effects of corporal punishment in a dog, a cat, or a llama. Punishment is to teach an animal when it has done wrong; it is thus we teach our domestic animals. We train a horse, we whip him when he has done wrong, we pat and encourage him when he does right. In the same way the Government trains us; it is like the bit in

28

the horse's mouth. There will always be horses who will try to get the bit between the teeth and run away, there will always be some who will jump the traces, rear and do various other things a well-trained horse ought not to do, and especially if there are any hereditary vicious traits perpetuated in him. When I say they have never been punished for so doing, I mean they have never been taught by physical or psychical punishment which varies in degree and nature with different individuals what we consider to be crime. We only know night by having day, we only know what is right by having wrong made perceptible to us as a contrast. In the same way those children who are taught to beg, to steal, to lie, to deceive, to be tricky, or cunning for the benefit of their parents, who are praised and caressed when they do wrong, these children grow up with the idea that evil is good. And after attaining manhood or womanhood, they carry into practice the training of their youth. The ignorant commiserate the father or mother, and say what a bad son or daughter, as the case may be.

Prof. Huxley shows the effect of education in his elementary physiology: "By the help of the brain we may acquire an infinity of artificial reflex actions, that is to say an action may require all our attention and all our volition for its first, second or third performance, but by frequent *repetition* it becomes, in a manner, part of our organization, and is performed without *volition* or even *consciousness.*" Acts whether good or bad become automatic and organic by repetition, which proves that forgiveness by a priest would not change the nature of a nervous centre.

As laws are made to govern human beings, and as they have so powerful an influence in determining the national character, they should, above all, be framed upon a thorough knowledge of man and woman, both physical and psychical. Laws must, of necessity, be one-sided which are formulated by men alone. There must be a combination of the two elements, otherwise the conflicting forces tending towards the destruction of the human race will never be adjusted.

The laws adduced from the philosophy of history, as the skeleton upon which to build any theory of politics, are considered indispensable. I maintain that the essential basis for any true Government is the thorough understanding of the embryonic development of individuals; and then comes that of tribes, societies and nations.

Under the present system of jurisprudence we

maintain an expensive judicial machinery to punish our morally imbecile whose diseased brains in the majority of cases are incurable. I used to say in my lectures that pauperism and crime are hereditary, that one thousand criminals had been traced back to one unfortunate, that the same names constantly reappear among the criminal and the pauper classes.

A judge sums up a case and cites precedents, similar cases in which the verdict was thus and thus. Precedent is a very unsafe guide when applied to individuals, the causes or determining influences of acts being rarely identical; there may be similarity, but on careful scientific analysis the differences may be great. The individual is not judged upon the sum of hereditary influences and education but on precedent. It is the philosophy of history and not the philosophy of the embryonic and individual history by which the person is judged. A judge decides that the State has no authority to prohibit the sale of liquor; he states he finds no precedent and it is contrary to law, and by this decision the prohibition law cannot be enforced. This decision is condemned when the criminal with his diseased brain, the offspring of the drunkard, is brought before the same

31

judge who again finds his precedent in committing the victim of this philosophy to prison.

In the humanitarian Government of the future, when our legislators have a thorough knowledge of psychology and pathology, our criminal courts will be presided over by a council of scientists who will examine into the nature and cause of the malady, whether the patient is curable or incurable, the effects of the environment and whether association with others of the same type by the power of suggestion would not intensify the malady instead of acting as a corrective. Those who are deemed incurable must be confined the same as insane and idiots are now. With the further development of scientific investigation into the causes of mental disease we will be able to master the conditions which favour its development.

It would be a fallacy to think that any law, any political institution, would be absolute in its remedial effects in every country, among all peoples, or even among every class of a community. When we realize how the characters of individuals vary, we understand how necessary is a thorough knowledge of psychology to legislate properly for the minds of men. The same prescription as a remedy for a specific disease is not efficacious with all persons alike; the physician must watch the cause of disease and judge in each individual case. He cannot state absolutely what the result in every case will be, but he may say that this course of treatment has proved efficacious in nine cases out of ten.

To acquire a thorough knowledge of psychology one must study physiology.

The spirit part of us is not something distinct from the material, something shadowy, undefinable; wafted down from no one knows where, to inhabit the body of each new-born child; then wafted back again to no one knows where. The psychical part, the spirit, develops with the development of the physical structure and disintegrates with the disintegration of structure.

To become conscious of a thing is to know a thing, to become aware of it, to have an idea of it. When we are asleep we cannot be said to be conscious of a thing.

Anæsthetics, an accident, disease, may render us unconscious; in these instances the spirit does not leave the body to follow its own devices, but still it is oblivious of what is going on in the body, or what is being done to the body. Hence, it follows that which affected the material, affected its properties, the spiritual. The psychical is simply the expression of the physical constitution.

It is generally accepted by scientists that consciousness is an acquisition of nervous tissue. Consciousness is the sum of the education of the nervous centres. These nervous centres have become educated by experience; they demonstrate their education when stimulated. The organization of the nervous centre is a physical fact, its psychical expression depends upon the physical constitution.

If a nervous or motor centre becomes degenerated through disease causing failure of nutrition, the movements or psychical acts, of which it forms a part, become disorganized. It is clearly shown how small a part consciousness plays in our habitual actions. If deeply engaged in conversation, that is if our higher nervous centres are engrossed in their own activity, we may perform several acts of which we are unconscious; if out walking, a lady may cross streets, avoid obstacles, get out of the way of puddles, lift her dress, and be entirely oblivious of the several acts unless her attention be especially directed to them.

The study of epilepsy abounds with instances of

automatic acts continued in states of unconsciousness, called post-epileptic automatic acts.

Physiology teaches us that the property of nervous tissue is to organize conscious acts into unconscious or automatic acts. In post-epileptic automatic acts, or post-hypnotic suggestions, the power which the higher centres have of controlling these automatic acts is for the time suspended or rendered inactive. Many diseases originating in different parts of the body may have such physical effects upon the brain as to render the will impotent.

Physical and moral well-being are interdependent. Any portion of the brain becoming hypertrophied or atrophied through disease will produce its corresponding psychical effect. Since the investigations of Ferrier and others with regard to the localization of sensory and motor centres of the brain, great advances have been made in positive psychology. Irritation of particular motor centres will produce certain movements. Irritation of sensory centres by disease, or a certain group of cells becoming hyperexcitable, often produce auditory or ocular spectra, or other hallucinations, which often give rise to uncontrollable impulses. Micro-organisms introduced into the body may produce morbid effects of which we have no knowledge.

As long as disease is restricted to a part of the body which has only a limited effect upon the brain, it may not affect the psychical, the moral well-being, but if not so restricted, again, it may affect the higher cerebral centres, and make us cognizant of its destructive power in criminal and immoral acts.

Although the anatomy of the brain shows that the cortex is imperfectly developed in children, that the fibres and cells are not yet differentiated, still we see we have not entirely to do with virgin elements in the precocious acts of children. A predisposition, a latent power to respond to particular stimuli, is there.

We often have thefts and other crimes performed by very young children showing abnormal conditions of the brain. Supposing as a child grows to manhood or womanhood there is arrest of development of particular nervous centres or groups of cells. The apparent man or woman may still be as irresponsible as the child, as far as the will, conscience, reason, determine our actions.

Several scientists have lately given their attention to arrest of development during the embryonic period, causing a particular portion of the brain to retain the

3 *

36

form of some primitive type, whereas the outward man may develop, to all external appearances, the same as other individuals. Gegenbaur calls it the reappearance of a more primitive organization, or a reversion to a primary state.

I used to say in my lectures the beast pursues the bent of his own nature, and he is contented or uneasy as this bent finds satisfaction or is deprived of it. No one considers it a crime for a tiger to devour a man. It is his nature thus to do; nature has made him so. How much more is it a crime in the human tiger?

All these evidences of imperfection in human character—all the evidences that the bloodthirstiness of the tiger is not yet burned out of matter, or that the cunning of the fox will manifest itself in man, when the fox is made the basis of his character—go to show how careful all should be who assume the responsibility of adding to the population of the world.

With idiots, or very pronounced deviations from the normal type, we realize something is wrong, and act accordingly; but where it does not come within range of our philosophy we look upon them as bad, if not wicked.

What effects may not the toxic products of a

diseased mother have in modifying the coenæsthesis of the developing embryo? Of course, we know that nature guards against this by the antiseptic properties of the metabolism of the body at this period, but in the abnormal conditions of disease the metabolic products may be different.

When a part becomes functionally active there is an increased blood supply to the part. The brain is the most active part of the organism, therefore is the most richly supplied with blood. Thinking, willing, in fact, consciousness, are functional activities of the cerebral hemispheres. When the brain is active the stimulus is communicated to the arteries, causing them to dilate, and the flow of blood is increased.

The proof that the functional activity of the cerebral hemispheres is dependent upon the supply and quality of the blood is that when the flow of blood is interrupted or lessened, thought becomes almost impossible, or there is total unconsciousness. If the cells of the cortex do not respond to stimuli when there is less quantity of blood, or the blood is not in its normal condition, it shows that there is a direct relation between the blood and mental states. When the circulation is increased beyond a certain point, delirium is produced, as is seen in fevers. This undoubtedly arises from the extra tissue change. Do voluntary acts involve molecular changes—metabolism of the higher nervous centres? This may account for the fact that nature has been economical and made the greater number of our acts automatic, as physiologists maintain.

The quality of our mind depends upon the physical condition of the cerebral substance. If the blood is diseased, it no longer furnishes proper nutriment, and there frequently follows morbid vascular conditions; schlchrosis, cysts, tumours are found in different parts, and there is often softening of the medullary substance.

There is reason to suppose that there is increased metabolism after work within certain limits; but suppose the blood has not those chemical constituents which the part demands? A part becomes strengthened and grows by exercise owing to the increased blood supply; if the blood is diseased it no longer furnishes the proper nutriment, and there follows degeneracy.

Degeneration of the frontal lobes causes dementia, the temporal lobes deafness; according to Ferrier, lesion of the occipital lobes will cause blindness,

39

degeneration or lesion of Broca's convolution and the Island of Reil is followed by aphasia, degeneration, hæmorrhage, or lesion, of motor centres; causes paralysis, or otherwise disorganizing co-ordinated movements.

It has been insisted upon by many that a great many fallacies which obstruct our path to exact knowledge have arisen from personifying abstractions. Perhaps the greatest fallacy is the personifying of conscience—his conscience pricks him—my conscience condemns me.

Conscience is the sum of the modifications of our higher nervous centres. Without the cerebral hemispheres a man would have no conscience or even consciousness. The conscience is determined by the education and development of the cortical cells, its organized experience. Therefore it is a fallacy to judge the consciences of nations or of individuals by our own conscience. Their nervous cells have not been educated in the same manner as ours. If we wish to understand an action we must ask ourselves what experiences have been organized into the cerebral substance of the subject, also the inherited organized experiences and the physical state, whether in an abnormal or normal condition; to these might

40

be added motive, and then we get a fair idea of the conscience of the individual.

Will, free will, like consciousness, is a property of this cerebral hemisphere; we may be conscious of a thing or particular state and yet may not be able to control or to alter it. Will is strengthened by use. The will represents so much energy stored up in the higher centres of the brain, the more it is exercised the more its capital is increased. Will is the power which we have to control or excite educated nervous centres into activity. It is not independent of these centres, it is only a stimulus acting upon them. When the nervous centres become diseased, degenerated, the will has no more power over them.

Weakness or malformation of an organ of the body involves a corresponding inaction of function. We perceive how useless it is to expect good, useful citizens from diseased bodies. A Government should look after the internal welfare of its people, and these will repay the nation by being healthy, useful citizens instead of being, as now, so terrible a blotch upon civilization.

The nation which produced the highest type of man and woman had in it the best germ of a true Government. A nation which appears the most powerful, yet has the greatest number of half-witted, of paupers, and of criminals as concomitants, is not the best governed. On the contrary, the seeds of decay have already taken root. If the Government is the representative of the people, the better the people the better the Government; conversely, the better the Government the better the people must follow as a logical sequence. As Socrates said, "The man who does nothing well is neither useful nor agreeable to the gods," and also said, "When an artisan goes wrong it is usually from pure ignorance or incapacity. He is willing to do good work if he is able." To this I would supplement by saying, if a man is born with good propensities he is willing to be good; and if not, he cannot help himself if education is not brought to bear to neutralize the evil tendencies.

A man who has studied at a school of agriculture and has also made practical application of his knowledge makes the best farmer. The laws of a State are not made to govern the vegetable world, nor to control the actions of the birds of the air, nor the beasts of the field, but they are made to restrain and control the actions of human beings. Very well, then;

men and women who have a thorough knowledge of psychology and its bearings upon political economy will make the best law-givers.

Some sociologists have maintained that this knowledge is not necessary—that spiritual and temporal power should be kept distinct. This theory is delusive. What is temporal power for? To govern the spiritual.

The Government is the physician for the evils afflicting its people. The morbid products of disease generated in poverty, misery and ignorance are to be prescribed for. The hereditary effects from shattered nervous systems, drunkenness, pauperism, prostitution, are already at work. What is the remedy? We cannot shoot the people down, and thereby eradicate the cause, but we can bring to bear influences which will counteract the evils already existing, and at the same time bring to bear scientific laws to eliminate the cause in succeeding generations.

Utility is the plea for every law and every institution. We provide prisons and other asylums for the insane, the inebriate, and the foundling, because it is utilitarian to do so. It is for the public good to place some kind of restraint over these manifestations and not let them be at large; more especially that it may have a deterring influence on others. How

43

much more utilitarian is it to carry into effect some scheme of supervision which may prevent instead of cure!

Utility is said to be the initiator of justice. What may be justice to one may be injustice to another ; so, in the government of a nation, an action must be judged from the standard of utility rather than from an idea of justice to the individual. A man commits a crime. From a utilitarian view that man should be punished for the safety and the interest of society. From the criteria of justice he may not be responsible for his actions; he may have a malformation of the brain or hereditary instincts over which he has no control. In this case what is the highest form of utility ? What is the most perfect justice ? The highest form of utility is to try to eradicate those conditions which make criminals possible, and, if we cannot do this, then to make such laws that it will be almost impossible to perpetuate this diseased condition of society. The most perfect justice would be so to legislate as never to have such persons born.

Is it the greatest utility to increase the inland revenue by encouraging the liquor traffic at the expense of making the standard of citizens poor in

44

neurotics, inebriates, epileptics, &c.? Is utility not better realized by having a race of fine, healthy men and women, who raise the standard of labour, and by probity, valour and morality enrich the coffers of a nation a thousand-fold?

Compare the amount derived from the tariff on liquor with the expense of supporting houses of correction, poor-houses and various asylums resulting therefrom, and see whether the Government is gainer or loser by its one-sided view of political economy. But when these institutions are considered as necessary adjuncts to any well-organized society, it proves that there must be some serious miscarriage of justice, some dangerous misapprehension of the meaning of utility.

Organized energy of many has greater power than individual energy. When this power is directed for good, it must of necessity have greater results than the efforts of a single individual working in a given direction. We cannot estimate exactly the force required to move an object until we have made tentative trials; the larger and heavier the body the greater the force required.

Although it may appear impossible to control the enormous number of incompetent individuals who need their energies directed in the right channel, it is really not so. All that is necessary is the organization of available material which will generate a powerful influence for good.

Suppose there were a combination of force for good, equal to the police force of any city. Suppose there were a standing army directing its force for producing healthy conditions in which people would come into the world properly.

The function of a good Government, it is said, is to interfere as little as possible with individual liberty; the province of Government is defence and nothing else. This interpretation is misleading. The law defends one human being against another. This is termed right, but even in doing this it is performing a two-fold function. The law insists upon individual rights being observed; when this is violated by an assault it punishes the offender, and therefore teaches self-control. By certainty of punishment it deters others from committing a like offence. This proves that in its fundamental principle the law educates as well as governs.

Is Government simply for defence, and has that been its sole office? By no means; it has developed the character of its subjects by enforcing obedience to certain civilizing rules. The well-to-do classes have other means by which they are humanized—education and association of ennobling ideas. What have the poor? Absolutely nothing except the law. Supposing Anarchists had their way, and the Government could place no restraint on its subjects, what would be the result? Withdraw the only civilizing power, and anarchy reigns supreme. Liberty differs from licence in being the co-operation of the many to maintain order.

The most absolute anarchy was when prehistoric man had absolute freedom of action and maintained it at the risk of his life. The more enlightened and perfect human beings are, the more they will respect the rights of others. I have awakened to the fact that the sovereignty of the individual if carried to the extreme is a pernicious doctrine. If a man has contracted some terrible, contagious disease, and is unable to help or doctor himself, would it be right for us to say: "You had perfect liberty of action, therefore you have caught the disease and you may die alone"? No, the true interests of Government would be to render what aid the community had to offer. So with a large majority of individuals, they are unable and incapable

47

of helping themselves; they are suffering from what, if left to itself, is an incurable disease of poverty, incompetency, laziness or inertia in the organism.

It is useless to talk of individual rights in these cases, unless it be the rights of invalids to be prescribed for. The solution is not in doing away with Government or limiting the power of Government, but in having a good Government which will be no restraint to those who recognize the necessity of conforming to certain laws if they wish to live together as civilized beings. Those who do not realize this must be taught by a good Government recognizing the rights of humanity.

It is a great mistake to say over-population; we should say bad population. There are not sufficient superior people to give even higher tone to life, thought and actions; and there are too many below a very ordinary standard of physical and mental development. And it is among the latter class that the birth-rate is so high. They would double their population in a very short time if their children survived the deplorable conditions of ignorance and poverty in which they are born. Yet what a commentary on civilization is it that our political economists have to say that although the birth-rate is so high among these wretched, incapable human beings, it is a blessing for humanity that the death-rate also is very large, as that balances, to some extent, the evil. No other remedy can be found ! ! ! except that the mortality is so great ! ! !

Is it nothing to bring human beings into the world in ignorance to suffer? Or do they belong to a class who have no feelings? It is simply begging the question to say that by a kind ordinance of nature those who will not work (or do not know how) will have to suffer by becoming paupers, or will be reduced to destitution. They are not the only ones who suffer; they crowd out the honest wage-earner or become parasites on those who do earn their livelihood. Are we right in saying that this class must always represent the destitute and miserable?

We say to these poor wretches: "You ought to be good, honest, healthy men and women." They turn and say to us: "How ought we to be healthy? We have never been taught anything which conduces to health; we know nothing about morality." For example, were I to ask a plumber to paint a house he would say to me: "I do not know how to paint; I never learned the trade." Which would be the most to blame—I who order him to do something about which he knows nothing, or the plumber because he is ignorant?

Are the half-developed, diseased, almost idiotic, hungry wretches skulking about our highways and byways responsible for their condition or ignorance? Is it not, rather, the fault of those who have the power and intelligence, and yet neglect to employ either for their instruction? They are but the creatures of circumstances, governed by hereditary instincts for evil generated in ignorance, by associated evil habits, by insufficient food, and foul air causing degeneracy of the system, by too easy access to liquor, having no moral restraint except the law, no aim, no systematic training or education conducing to physical and mental development, living in bad dwellings. What wonder that the consequence is increase of immorality and wretched population, their environments being entirely for evil, nothing for good.

A humanitarian Government would stigmatize the marriages of the unfit as crimes; it would legislate to prevent the birth of the criminal rather than legislate to punish him after he is born. I used to say in my lectures that a person with deformed legs will walk with an imperfect gait. But if deformed bodies determine ungainly physical action, why should not a deformed brain determine inconsistent mental action? The character both of individual thought and of action is almost dependent upon the condition of the human organs through which they are evolved.

A man is unfortunate enough to have some terrible disease; he desires to marry; if he has no deterring influence to counteract this wish he satisfies the desire. But if he were confronted with such consequences as these—I shall be ostracized by an educated public opinion if I reproduce my diseased condition in my offspring, if I produce a criminal I shall be adjudged the culprit—it would make him reflect. The humanitarian Government by stigmatizing such marriages as crimes would gradually enforce upon the public mind the responsibility of parentage.

The effect that the law has upon popular opinion cannot be more clearly demonstrated than in the following opprobrious epithets: He is a bastard—he was a *natural* child (hush! don't speak of it under the breath)—he is a *love* child—he is *illegitimate*. Even a woman's virtue is decided by the law in the public mind. We see this with those who wish to ruin a woman's reputation in many sneers and insinuations which have their origin in the presence or absence of legal formalities. So deeply ingrained is the idea of marriage become synonymous with the legal ceremony in the public mind that it is difficult to disassociate the two. We see what constitutes the virtue in the following, in the due exercise of *lawful* love; but *illicit* love is *sin*. Lawful is derived from *legalis lex legis*—law, lawful, created by the law, therefore right, virtuous; illicit—*il*, not; and *liceo*, to be allowable; illicit is sin because it is not licensed. A man who sells liquor without a licence to demoralize a community is bad, a criminal; but if he does the same thing to destroy his fellow-creatures with a licence he is a law-abiding citizen, therefore respectable. Even more we see its effect when she is ruined, dishonoured, and the pleadings of the victim to give her back her honour, to make reparation. Public opinion is an extraordinary conglomeration. It is the law which compels the man to the observance of certain civilizing rules, it opposes the human man to the animal man; otherwise he is apt to love and ride away. The law is the regulator of an animal passion and enforces obedience either directly by legal punishment or indirectly by the effect of the stigma in social ostracism. As our ideas become more definite with regard to what is right and wrong, we will use every influence that the law, religion or society, can do to intensify the good and stigmatize the evil.

It is true that very imperfectly organized individuals may be surrounded by such superior influences as will call forth only good acts or thoughts; but a change of influences merely is required to develop latent qualities. The method of human improvement by the action of better influences and examples is palliative merely, while a radical change must proceed from scientific propagation.

Hence it will be at once perceived that every human effort must primarily result from some inherent capacity being acted upon by some exterior force. The adult has just such a conscience as the circumstances of his birth and succeeding education and surroundings have shaped for him. Every person is just what he is made to be by his parentage and education.

But to make this fact still more pointed and clear, let this act be whatever it may, from the slightest offence up to murder, it would be committed by any other person provided that person had been conceived, gestated, grown, and surrounded by the same circumstances as the one who committed the act, since such a person would have been an exact counterpart of the other.

Again, it is to be observed that the consciences of various individuals are as different as their other points of character, while each individual proceeds to adjudge every other person by his own conscience. This is manifestly unjust. Even the Christian Master's teachings in this regard were entirely different from the practice of His professed followers. He said, "Judge not." Neither can we judge justly of the act of any other person. We must first put ourselves in his place; not merely conceive that we are in it, but actually place ourselves in the same personal condition in which the person to be judged was in when the act was committed upon which judgment is to be passed. The utter impossibility of this at once demonstrates the like impossibility of a just judgment being rendered in any case whatever.

No person living can stand amid the people and say, and say it truly, "I alone am responsible for what I do." No one can be completely divorced from things around so as to make this possible. In the first place, everyone is a natural result of the conditions in which he or she was produced; and in the

54

second place, this aggregate of results is open to the influence of everything by which it is surrounded, and is made to act in given directions, in proportion as those influences are strong or weak in those directions. No one is his own master or her own mistress, but on the contrary, persons are the slaves of the influences in which they live, conjoined with their inherited tendencies.

The larger part of what is called crime is the result of hereditary instincts and habits engendered by pernicious environments. The crime due to hereditary criminal instincts—how is one to punish or judge it with any idea of justice? The true criminals in this case are the parents.

Again, with the irresponsible actions of the so-called criminal, labouring under the delusions of mad delirium, of uncontrollable impulses, the victim of circumstances over which he has had no control—how do we expect to punish him with any idea of justice? I used to say this class are fit subjects for hospitals, and the jailors should be physicians. I said in 1870 that our prisons should be turned into vast reformatory workshops, from which the unfortunate may emerge to be useful citizens, instead of the alienated citizens they now are.

55

If a man produces a pauper, an idiot, a drunkard, a criminal, an insane being, he is interfering with other people's rights; he is producing the "sweater," in the shape of incompetent labour, the incapacitated to be supported by the capable. It is the improvident living at the expense of the provident. And it is these false ideas of liberty which make the struggle for existence so terrible.

The day is not far distant when the healthy, strong wage-earner will rise in rebellion and say: "You have no right to throw your burden upon my shoulders."

Society of human beings is for mutual benefit, and if humanity wishes to derive that benefit it must conform to certain laws of interdependence, otherwise it is simply anarchy. It is like one part of the body saying to another part: "I shall not work with you any longer. I am at perfect liberty to become diseased and to propagate diseased cells if I choose." It is so, but the whole body suffers. So with the body social. Each member thinks that his or her little impetus in a given direction cannot influence the great mass, forgetting that it is the sum of these influences which constitutes organized life.

A humanitarian Government would not grant a marriage licence to anyone malformed or having a transmissible or communicable disease. No marriage certificate would be given to the contracting parties unless they understood physiology and had some visible means of subsistence. The lessons in physiology would be given gratis, if necessary, by the medical officials of the Government.

Insertion of the banns of marriage might take the following form in an Official Gazette: J. S. wishes to contract a marriage with A. B., both contracting parties deeming themselves worthy to become parents —the pedigree of J. S., occupation, &c.—The same formula with regard to A. B. It would give the community at large a fair idea of the *fitness* of the contracting parties.

These laws would be provisory until human beings were educated to the standard that they would never think of becoming parents if they were unworthy. When the birth of a child is registered, a description of occupation and physical condition of both parents should be given and certified by physicians under the pay of the Government, a copy of this registration to be given to anyone. In after years this certificate should be produced before any marriage certificate could be obtained, thus insuring mutual benefit for all.

The quality of the blood of a people is of national

importance. It represents its strength in useful, noble, intelligent citizens; it reveals its weakness in disease, debauchery, and pauperism. People are not born equal. No two individuals are absolutely equal. Some are born kings: others are born slaves—slaves to hereditary diseases, propensities, appetites, and shattered nervous systems. The kings are those who have healthy, perfect forms, highly developed brains, which by the richness of blood can reason well, whose senses are more highly developed. They are, indeed, nature's noblemen. An aristocracy which knows no equal!

Is the man whose retina has become insensible through disease equal to him with highly sensitive retina? This physical defect will often delude his judgment with regard to objective phenomena. Is the man whose brain has become diseased through the abuse of alcohol, or he who has contracted some disease with the result of having little nodes formed in his brain interfering with the mechanism, or he with hyperexcitability of certain nervous centres interfering with the nutrition of other parts, equal to him with highly developed healthy brain? In what does value consist?

A jewel may be of more worth than a stone a

million times the size; the value here is not in quantity, but in quality. So it is the quality of our intellect, of our physical development which stamps our rank and makes us either inferior or superior to our fellow-creatures.

A man now can boast that he has an ancestral estate and a title, because he happens to be the eldest son, and with these he very often inherits some physical defect or terrible disease of the blood which makes him a representative of bad blood instead of blue blood. Very often he has no ancestral estate or any other possessions, but a title only, and he sells that birthright for a mess of pottage.

Science has shown that the eldest child is rarely so well developed either mentally or physically as the succeeding children. The maternal organs become more developed in successive pregnancies, and the embryo derives the benefit. There is a plurality of causes acting for and against higher development, so each individual must be judged personally by the result.

" Pride, sense of dignity, and self-respect are very conspicuously exhibited by well-treated dogs. As with man so with the friend of man, it is only those whose lines of fortune have fallen in pleasant places, and whose feelings may therefore be said to have

59

profited by the refining influences of culture, that display in any conspicuous measure the emotions in question. 'Curs of low degree,' and even many dogs of better social position, have never enjoyed those conditions essential to moral refinement, which alone can engender a true sense of self-respect and dignity. A 'low-life' dog may not like to have his tail pulled, any more than a gutter child may like to have his ears boxed; but here it is physical pain rather than wounded pride that causes the smart. Among 'high-life' dogs, however, the case is different. Here wounded sensibilities and loss of esteem are capable of producing much keener suffering than is mere physical pain; so that among such dogs a whipping produces quite a different and a much more lasting effect than in the case of their rougher brethren, who, as soon as it is over, give themselves a shake, and think no more about it."*

Those who have never enjoyed those conditions essential to moral refinement, which *alone* can engender a true sense of self-respect and dignity. How many of our fellow-creatures can this be applied to? How many are surrounded by conditions which conduce to self-respect and dignity? When

* Romane's "Animal Intelligence."

we speak in contemptuous terms of the populace, the rabble, the mob, we are lowering the self-respect and dignity of humanity. The aim of humanitarian Government would be to develop these refined sensitive feelings in the majority. Under our present system we say—What did he rise from ? he came from the gutter ; not—Is he in himself as good, noble, or superior? We scandalize, we defame, until self-respect is obliterated. We often hear: Be careful in your dealings with So-and-so, he or she has nothing to lose. The sooner we extend this responsibility and give the poor something to lose, the sooner we shall break down false barriers and raise the standard of humanity.

To give an example how people should be designated by their rightful titles I may cite the story of Beethoven and his brother. The latter to distinguish himself from his shiftless brother signed himself—"Von Beethoven, land owner." The composer in contradistinction signed himself—"Beethoven, brain owner."

A humanitarian Government would give titles only as life endowments. Prince or Princess being the highest title, it would be the most difficult to acquire; the genealogical tree would have to be

61

proven healthy for at least three generations; his or her pedigree would also have to prove that there is no taint of insanity, crime, drunkenness, or transmissible disease; it would be absolutely essential that he or she should have attained a certain standard of physical and psychical development. Any physical or psychical defect excludes human beings from the aristocracy of humanity. The aim of a humanitarian Government in creating such an aristocracy would be to counteract the evil effects of a plutocracy, and give an impetus to physical and psychical development.

A humanitarian Government would insist that newspapers, in reporting crimes, should frame the record in language that would present a moral lesson to those who read, instead of permitting such accounts to demoralize the minds of their readers. It seems terrible to have to say that moral progress does not keep pace with intellectual, and this is one of the reasons.

The success of our sensational papers depends upon how vividly the scene is described, to excite and work upon the imagination, to make the readers *feel* the thing has actually occurred or is occurring, to make them live through the experience. And if some tale of horror is pored over by the reader with suspended breath, hair on end, the body in a perspiration, the success of the imaginary writer is complete. And why? Because the same nervous process is stimulated for the precept and the image, the presentation and representation, only the reality may be more intense. By repetition these emotions become organic, become predisposed to respond to this kind of stimulus. When we read, or are told, a pathetic story, or see a pathetic scene, we feel sad and *weep* because the same nerves and muscles would be affected if it were happening to us; and the more intensely we would feel it, the more intensely we sympathize. In the same way a funny story makes us *laugh*, which will vary in individuals according to the *sense* of the ludicrous. Again, the *anger* we experience when we read of an injustice done to some one: the blood boils, so to speak, according to the intensity of the individual, as if the injustice were personal; the physical effects are the same, only weaker. People very often will be able to repeat a choice bit of scandal, they will be able to relate every circumstance, dates, and most ordinary details connected with it, so vivid, so intense are the impressions made upon the nervous centres by anything which appeals to our passions. These same

63

people can scarcely repeat any noble deed or generous action they have read about, or if they do repeat the circumstance, it is soon forgotten. It is because that which is humane, noble, pure, refined, is only acquired and retained by the greatest effort.

The literature of a nation has a powerful influence on the morale of its people.

We read of a disgusting scene or a horrible crime, and our imagination fills in the details. It may be only a passing breath of contamination, the effect of which depends upon the mind through which it is communicated, but the more the same process is repeated the greater will be its influence for evil. The scenes being actually witnessed, the other senses may rise in revolt and counteract the evil effect. The actual squalor, smell, foul language, would shock the senses, producing a sensation of disgust which operates against the evil effects on the imagination of the observer. In seeing the actuality we realize the terrible conditions which have destroyed the nobler instincts of our less fortunate fellow-creatures, and the event does not steal like a thief in the night to rob us of our soul's greatest treasure, a pure mind.

Should we be justified in giving a description of the qualities of sewage by saying that it was true to nature, and therefore you can make no objection? But is sewage a natural production, or is it not the refuse of the filth of the inhabitants of a city? The disgusting details of ignorance and debauchery permeating our so-called realistic literature; is it true to nature or is it the refuse of the filth of a diseased condition of society?

These so-called realistic novels, of what are they realistic? Of nature in all her purity, of the natural instincts of the animal kingdom? No, they are a portraiture of the morbid, diseased misconceptions of the human body. Could such scenes be depicted of the brutes? It is humanity which has sunk so low that its habits and thoughts are an outrage on all nature. "For such as are thy habitual thoughts, such also will be the character of thy mind."

In opposition to realism is idealism, the effect of which is to elevate the mind. No one can hope to reach the ideal, but everyone can strive after that higher life. In so doing the soul unconsciously reflects the model. If an artist is going to learn his art, deformities are not placed as models before him, nor daubs from which he is to receive his idea of colouring, but the finest examples to be found; nor is he satisfied until he has studied the greatest masters, whose souls are reflected upon inspired canvases, whence his own soul becomes imbued until the very tone of a Raphael speaks through him to mankind.

Humanitarian Government would have the science of sociology taught in schools. It would teach the young what diseases they are most subject to; for instance, gastric, lung, or nervous complaints. A certificate would be given by physicians under governmental pay whether the children have weak eyes or other degeneration of organs, suggesting for what kind of work they are most suited. It would not interfere with individual liberty, it would only utilize the skill of those who have devoted time and labour to acquire this knowledge for the benefit of those who have neither, and acquire it after bitter experience, when very often it is too late to be of any practical benefit. These certificates are granted for *mental* fitness, so should be for physical.

These physicians would teach the working poor how best to avoid conditions which would bring on disease; their own strength or weakness; they would teach them to only adapt themselves to those conditions for which they are physically capable. The practical aim of such measures would be to extend the knowledge of physiology.

In this body social there must be division of labour,

and by this division they expend their energy in different kinds of knowledge. The capital of experiences is invested in different acquisitions. Some are portioned off to do manual labour, others mental. The energy of a human body is limited. Those who expend their energy in doing manual labour have not that energy left for mental labour. A woman or man's value to a community is his *usefulness*, and if he has no value he is a parasite.

In a humanitarian Government the cabinet would be composed of philosophers, representing every branch of science. They would utilize this knowledge and would feel how necessary it is to look after the internal welfare of the people. They would see that every building were properly ventilated and in a sanitary condition where working women or men are employed. They would see that the very poor had sanitary dwellings. They would erect cheap lodging-houses for improvident women, and utilize the unemployed labour in every district.

A humanitarian Government would feel how necessary it is to erect in each district, just as police-stations are now, buildings containing large halls and pulpits. Each would have a staff of officers, but they would be there to report upon and deal with causes instead of effects. This staff would be composed of trained professors in sociology, mechanics and nurses. They would be able to teach the laws of heredity, the result of acquired habits, and how these react upon the physical and moral condition of offspring.

There should be female philosophers to teach mothers the full responsibility of maternity.

The mechanics would impart the latest improvements in different types of work; how time and labour could be saved by adopting this or that method. Their efforts would raise the standard of labour.

These officers could look after the domestic policy of each district, report cases of dire necessity and organize a system for temporary relief in urgent cases, not as charity, but as loans to be repaid by the recipients.

The inhabitants of a district would be able to lodge complaints, and suggest improvements.

The purposes of these buildings would be to educate the masses by a system of oral education. Our wage-earners have but little time for reading; the age for them is past for studious application, but it requires very little effort to listen. After a hard day's work of continued bodily exertion they have no energy left for much brain exertion.

It would be the same system as when the masses were taught in the market-place in Athens by

Demosthenes, or when they stood and listened to the Sermon on the Mount. In primitive ages all knowledge was given publicity by word of mouth, handed down from father to son.

The great fault of the present age is that knowledge is disseminated by books which instruct the educated, while the poor and ignorant are instructed by demagogues or by inferior and sensational literature which develops depraved appetites. In ancient times people travelled for days to converse with some known sage to learn what truths he had to offer. The populace were taught orally then, and must be so taught now, for the masses leave school at a comparatively early age, and have very little understanding of their own organisms.

The specific aim of humanitarian Government would be to concentrate all that is noblest and divine in human nature, to strengthen and revivify the life-giving principle (protean matter) into greater and more perfect types of human beings.

Chapter 7
The Rapid Multiplication of the Unfit
Introduced by Michael W. Perry

> *But in any attempt to raise the standard of humanity, to aid evolution, we must take into consideration that it is not the survival of the fittest, but he survival of the unfit by means of their rapid multiplication in societies as presently organized.*

In this 1891 booklet Victoria Woodhull focused on the medical, social and economic factors that she believed promoted the "rapid multiplication of the unfit." Most of her arguments were based on conditions in England, where she lived, rather than the United States. There is even a hint that she believed in a "new country" where "sedentary occupations are rare" (8), was less likely be troubled with a rapidly multiplying unfit. Of course, as the United States industrialized and more people moved into cities, her arguments would apply.

Her medical arguments reference Michael Foster's *Text-book on Physiology*. Much of the medicine in that popular text, first published in 1877 (with numerous editions into 1890s), reflected a then-common belief that the human nervous system was like a watch that could get out of balance or a battery that could be depleted by stress. The result of either was illnesses, mental or physical, that would be passed in a Lamarckian sense along to children. Parents gave their children not simply the genes they had been born with but little copies of themselves influenced by everything that had happened in their lives. To give an extreme example, a woman who attempted an abortion might give birth to a child prenatally inclined toward murder. That belief muddled the distinction we now make between heredity and environment. People weren't either born *or* made, they were born *and* made.

> In the same way we build insane asylums to house our insane because they have lost their mental balance, so we build pauper institutions for those who have lost their physical balance. The vagabond, the pauper, is as much born and made one as the man of insane temperament under stress demonstrates his neurotic heredity and the criminal his pathological condition. The terrible effects on posterity of depleting our workers and then allowing them in ignorance to breed is the burning question for humanity. (8)

This meant that major social problems, from poverty to crime and mental illness, were the result of both hereditary and environment, with no clear distinction between the two. That was why Woodhull warned about "depleting our workers and then allowing them in ignorance to breed." Tired parents meant tired children. She took that argument even further and warned that future generations would not retain their "superior qualities" if the better members of the present generation were "overtaxed, overburdened, and under the strain." (13)

That was a dangerous argument because it claimed that burdening the healthy with the costs of caring for the 'unfit' would not only take idle pleasures from the lives of affluent taxpayers, but would create stresses that would turn at least some of the 'fit' into 'unfit.' Her hostility toward disabled children is particularly disturbing.

> In visiting one of these institutions a short time since, in one of the wards I saw a little child moving about with the aid of a chair, its body being too big and heavy for its legs; in another ward a nurse, who was carrying a baby covered with scrofulous sores, asked me if I would adopt it.... The doctor took great pride in showing me a child on whom he had just operated for hare-lip; my attention was drawn to the success he had in delivering a mother of an idiotic baby. What is the destiny of these children? They require able-bodied nurses from their birth, and able-bodied physicians to spend their valuable time over them.

What Woodhull was advancing is today called the 'quality of life' argument. It claims that society has no reason to keep alive those with major disabilities, that their lives are not worth their cost to society. These arguments almost inevitably surface when the debate turns to abortion or euthanasia, and in modern times it has been one of the chief points of contention between secular liberalism and traditional, Judeo-Christian religions. The reason is obvious. Secularism has trouble attaching value to individuals beyond their economic, artistic or intellectual value to society. The debate has been going on for over a century. You can read a portion of it in the *New Republic*, a liberal magazine, in a 1915 article entitled "The Control of Births."

> And to the mind of man it would mean a release from terror, and the adoption, openly and frankly, of the civilized creed that man must make himself the master of his fate; instead of natural selection and accident, human selection and reason; instead of a morality which is fear of punishment, a morality which is the making of a finer race. Fewer children and better ones is the only policy a modern state can afford."[1]

The *New Republic* did not state the case quite as "openly and frankly" as they claimed. Man was not to become "the master of his fate." Abstractions like that mean nothing. In practice, some men were to become the masters of the rest, to dictate "The Control of Births," and create what they considered a "finer race."

That contrasts to a Christmas pastoral New York City's Archbishop Patrick Hayes released in December of 1921. It defended the rights of poor mothers by turning to what was, for Catholics the supreme example of a good mother,

---

1. Michael W. Perry, ed. *The Pivot of Civilization in Historical Perspective* (Seattle: Inkling Books, 2001), 115. Quoted in *New Republic* (Mar. 6, 1915), 116.

writing: "The Christ-Child did not stay His own entrance into this mortal life because His mother was poor, roofless and without provision for the morrow.[2]

Even more disturbing is what happened when those same quality of life arguments were adopted in Germany and applied with Teutonic thoroughness. Dr. Leo Alexander, special medical consultant to the Nuremberg Trials, described the result in an article in the July 14, 1949 issue of *The New England Journal of Medicine*.

Even before the Nazis took open charge in Germany, a propaganda barrage was directed against the traditional compassionate nineteenth-century attitudes toward the chronically ill, and for the adoption of a utilitarian, Hegelian point of view. Sterilization and euthanasia of persons with chronic mental illnesses was discussed at a meeting of Bavarian psychiatrists in 1931....

It is rather significant that the German people were considered by their Nazi leaders more ready to accept the exterminations of the sick than those for political reasons. It was for that reason that the first exterminations of the latter group were carried out under the guise of sickness....

Whatever proportions these crimes finally assumed, it became evident to all who investigated them that they had started from small beginnings. The beginnings at first were merely a subtle shift in emphasis in the basic attitude of the physicians. It started with the acceptance of the attitude, basic in the euthanasia movement, that there is such a thing as life not worthy to be lived. This attitude in its early stages concerned itself merely with the severely and chronically sick. Gradually the sphere of those to be included in this category was enlarged to encompass the socially unproductive, the ideologically unwanted, the racially unwanted and finally all non-Germans. But it is important to realize that the infinitely small wedged-in lever from which this entire trend of mind received its impetus was the attitude toward the nonrehabilitable sick.

Fortunately, there was someone in this period who recognized what was happening in Germany and warned of where it might lead. In 1922, a well-known Catholic journalist and writer, G. K. Chesterton, published *Eugenics and Other Evils*. In an introduction entitled "To the Reader," he explained how his book came to be written. Notice how he could assume, as a matter of course, that eugenics was popular in the press, among "intellectuals," particularly those on the left such as playwright George Bernard Shaw (Chapter VI of *Pivot*), as well among the "ruling classes." The "scientific officialism and strict social organisation" that he criticized is virtually identical to that of Woodhull

---

2. Michael W. Perry, *The Pivot of Civilization in Historical Perspective* (Seattle: Inkling Books, 2001), 167. The archbishop was quoted in the *New York Times*, Dec. 18, 1921 on page 16. Two days later the paper published an angry response from Margaret Sanger, in which she claimed Catholics had no right "to make these ideas legislative acts." Sanger, deaf to the rights of those she disliked, assumed she could call for legal change while her opponents could not.

in "Humanitarian Government" or the *New Republic*'s state-directed policy of "human selection and reason."

Though most of the conclusions, especially towards the end, are conceived with reference to recent events, the actual bulk of preliminary notes about the science of Eugenics were written before the war. It was a time when this theme was the topic of the hour; when eugenic babies (not visibly very distinguishable from other babies) sprawled all over the illustrated papers; when the evolutionary fancy of Nietzsche was the new cry among the intellectuals; and when Mr. Bernard Shaw and others were considering the idea that to breed a man like a cart-horse was the true way to attain that higher civilisation of intellectual magnanimity and sympathetic insight which may be found in cart-horses. It may therefore appear that I took the opinion too controversially, and it seems to me that I sometimes took it too seriously. But the criticism of Eugenics soon expanded of itself into a more general criticism of a modern craze for scientific officialism and strict social organisation.

And then the hour came when I felt, not without relief, that I might well fling all my notes into the fire. The fire was a very big one, and was burning up bigger things than such pedantic quackeries. And, anyhow, the issue itself was being settled in a very different style. Scientific officialism and organisation in the State which had specialised in them, had gone to war with the older culture of Christendom. Either Prussianism would win and the protest would be hopeless, or Prussianism would lose and the protest would be needless. As the war advanced from poison gas to piracy against neutrals, it grew more and more plain that the scientifically organised State was not increasing in popularity. Whatever happened, no Englishmen would ever again go nosing round the stinks of that low laboratory, So I thought all I had written irrelevant, and put it out of my mind.

I am greatly grieved to say that it is not irrelevant. It has gradually grown apparent, to my astounded gaze, that the ruling classes in England are still proceeding on the assumption that Prussia is a pattern for the whole world. If parts of my book are nearly nine years old, most of their principles and proceedings are a great deal older. They can offer us nothing but the same stuffy science, the same bullying bureaucracy and the same terrorism by tenth-rate professors that have led the German Empire to its recent conspicuous triumph. For that reason, three years after the war with Prussia, I collect and publish these papers.[3]

---

3. G. K. Chesterton, *Eugenics and Other Evils* (Seattle: Inkling Books, 2000), 11. "Nine years old" would place his first remarks on eugenics in 1912–13, coinciding with the First International Congress on Eugenics in 1912 (see p. 13), when eugenics 'went mainstream' and was praised as a marvelous "new science" newspapers such as the *New York Times* (Chapter 12 in *Pivot*). Today many deny that eugenics was ever a science. But in its heyday it was regarded as such and opposition in the scientific community was rare and muted.

Unfortunately, that "same stuffy science," with the addition of raw racism, would lead Germany into yet another disastrous war.

We must now step back to Woodhull's earlier argument that the 'unfit' burden society and create stresses that turn even the children of the fit into unfit citizens. In chapter 12 of *Eugenics and Other Evils* (1922), G. K. Chesterton would describe the situation far more sympathetically than Woodhull, and he gave a traditional religious answer—if social evils are creating social ills, we should put an end to those social evils. Speaking of the callous capitalist who brutally exploits his workers, he wrote with eloquent sarcasm:

> However, he had made a mistake—as definite as a mistake in multiplication. It may be summarized thus: that the same inequality and insecurity that makes cheap labour may make bad labour, and at last no labour at all....
>
> .... And as time went on the terrible truth slowly declared itself; the degraded class really was degenerating. It was right and proper enough to use a man as a tool; but the tool, ceaselessly used, was being used up. It was quite reasonable and respectable, of course, to fling a man away like a tool; but when it was flung away in the rain the tool rusted. But the comparison to a tool was insufficient for an awful reason that had already begun to dawn upon the master's mind. If you pick up a hammer, you do not find a whole family of nails clinging to it. If you fling away a chisel by the roadside, it does not litter and leave a lot of little chisels. But the meanest of the tools, Man, had still this strange privilege which God had given him, doubtless by mistake. Despite all improvements in machinery, (the fittings technically described as "hands") were apparently growing worse. The firm was not only encumbered with one useless servant, but he immediately turned himself into five useless servants.[4]

For Chesterton this deterioration was a reason to treat the working poor better. But for those who supported eugenics, from Marxist intellectuals to wealthy capitalists (Woodhull had a foot in both camps), it was an argument to correct God's "mistake" by making sure the overburdened working poor had as few troublesome children as possible. Woodhull stated that solution bluntly, when she wrote: "For our plea to have an effect they must be given new nervous systems and healthy rich blood, in other words, they must not be bred." (16)

It's also important to realize that Woodhull and those who thought like her saw two sides to this problem. It wasn't just that the 'unfit' were having too many children, the 'fit' were also not having enough children, at least in relative terms. Here is how she described the latter problem. Notice how closely she links "fit" with the "better classes."

> ... High motives deter the fit from marrying until they are in a position to do so. Among the better classes marriage is deferred more and more, the standard of living is becoming higher among them, and more time is given

---

4. G. K. Chesterton, *Eugenics and Other Evils* (Seattle: Inkling Books, 2000), 88–89.

to education, whereas the unfit, who are not deterred by any qualms of conscience or apprehension of consequences, go on multiplying. And as the more highly developed are not perpetuated, or if perpetuated it is in fewer numbers, the thoughtless, improvident, degenerate, and diseased, multiply upon us. (17)

Those who want to follow how this argument, which typically used the term "foresight," developed over the years, as the 'fit' attempted to explain how their small families demonstrated their racial superiority, should see *Pivot*, particularly chapters II, III, VII, VIII and IX. Margaret Sanger's remarks are in chapter XXVI as well as in chapter 8 of the portion that republishes her 1922 best-seller, *The Pivot of Civilization*. There Sanger criticized those eugenists who wanted to force the 'fit' to have more children.

> ... But the scientific Eugenists fail to recognize that this restraint of fecundity is due to a deliberate foresight and is a conscious effort to elevate standards of living for the family and the children of the more responsible—and possibly more selfish—sections of the community. The appeal to enter again into competitive childbearing, for the benefit of the nation or the race, or any other abstraction, will fall on deaf ears.[5]

Wherever eugenics was discussed, racism lurked not far beneath the surface. We should not forget that Charles Darwin's *The Origin of Species* has as its subtitle: *By Means of Natural Selection or the Preservation of Favoured Races in the Struggle for Life*. That certain races and classes ought to "favoured" was one of eugenics' most deeply held beliefs, although it was typically discussed in coded language, so target populations would not be warned. And one of eugenics deepest fears was that modern life was out of balance, that "natural selection" and the "struggle for life" no longer "favoured" those who were truly the fittest.

That was certainly true of Woodhull. In the booklet that follows, she demonstrated that her remarks in *Humanitarian Government* about race not being a part of her thinking were a ruse. While she and those who thought like her might regard the more elite segments of another race more highly than their own 'unfit,' that did not mean they did not see some races as inherently inferior. Here is an illustration.

> A great many seem to think that interference with marriages of the unfit will only give great opportunities to races, lower on the scale of development who are multiplying so fast, to overcome and conquer the more advanced races. We have an example of this in the rapid multiplication of the negroes in America, who at some not far distant day will outnumber and overrun the whites if the rapid increase be not checked. Eventually, if America is owned and governed by negroes, would it be the survival of the fittest? The outlook is as ominous in Europe. (18)

---

5. Michael W. Perry, ed. *The Pivot of Civilization in Historical Perspective* (Seattle: Inkling Books, 2001), 234.

Her last remark may have referred to the immigration of poor Irish and Eastern Europeans (particularly Jews) into England. Some three decades later, in its November 1922 issue, Dr. Marie Stopes' *Birth Control News* would publish a letter from a reader, C. B. S. Mildmay, who would praise England's birth controllers for "lessening the number of births from inferior human parents." He went on to complain about "low-caste foreigners, especially from Eastern Europe, who were "like the rats," with "strong tribal instincts," and "compared with our people, they are cunning, bloodthirsty and cowardly." He noted that "very likely the Society [for Constructive Birth Control and Racial Progress] had this race question in hand." No doubt they did, that's precisely the sort of thing birth controllers intended to do. (His letter is republished in Appendix H of the Inkling edition of *Eugenics and Other Evils*.)

In addition to its class and racial bigotry, eugenics meant altering how society handled romance and marriage. Woodhull began a discussion of that when she wrote: "All artificial social inducements for the mating of unsuitable individuals are instruments for the multiplication of the unfit." (20) "Artificial" was her way of suggesting that our romantic life must be judged rather coldly by whether it improves the human race biologically. But if we drive love out of our culture and genes by selective breeding, have we really improved the human race? To his great credit, in 1912 one eugenist, David Starr Jordan of Stanford University, found that troubling.

> If they would, they must in time eliminate the most vital elements in human evolution—love and initiative. Love is the best basis for marriage, and love is a very real and noble thing, in spite of the baseness of many of its imitations.
>
> The value of eugenic study is in the diffusion of sound ideas about life and parenthood. Government can do something by refusing parenthood to those who cannot care for themselves because of feeblemindedness, disease and vice, but legislation must be undertaken very cautiously, giving the individual the benefit of all doubt.[6]

That good sense brought upon its author the wrath of the *New York Times*, which rose up and questioned how a "well-informed scientist" could make such remarks. "No eugenist worth of the name" it claimed, "dreams of applying to the human race any methods of selection even remotely resembling those of the so-called 'plant-wizard.' [Luther Burbank]"

Unfortunately, Woodhull, whose three marriages seem to have been at least in part about getting ahead financially and socially, was more calculating than Professor Jordan. To get around the obvious crudity of scientific mating and lend it a mystical air, she suggested making "a religion of the procreative prin-

---

6. Michael W. Perry, *The Pivot of Civilization in Historical Perspective* (Seattle: Inkling Books, 2001), 75. Quoting from a front page story in the September 1, 1921 issue of the *New York Times*. The paper's anger at him was published on September 3 on page 10.

ciple. Our girls and boys must be taught how sacred is the life-giving principle." (39) The idea made no sense. Eugenics, a cold, distant and snobbish ideology, is not going to supplant the many attractions between men and women that have existed for generations beyond counting.

The same is true of religion. To create a eugenic religion, the traditional, well-established religions that teach their followers to "be fruitful and multiply," as well as to value all human life, would have to be brutally suppressed or at least marginalized. That eugenists have been unable to do, despite their best efforts. Given free scope, a faith that does not regard people as mere animals and that doesn't plan to weed out the 'unfit' is likely to prove far more popular with ordinary people than one that does.

The failure to create a religion around the "procreative principle" is why, a century and more after Woodhull's remarks, eugenics operates in deep cover. It has no illusion that it will ever become popular outside a few narrow circles. Instead, it works to persuade those it regards as biological and social nuisances to focus on sex and not burden their lives with an unwanted child. Virtually all attempts to enlist people in the once grand cause have disappeared.

THE RAPID MULTIPLICATION

OF

THE UNFIT.

BY

VICTORIA C. WOODHULL MARTIN.

17, HYDE PARK GATE, LONDON;
AND
142, WEST 70th STREET, NEW YORK CITY, U.S.A.
1891.

THE RAPID MULTIPLICATION
OF
THE UNFIT.

One of the most fruitful sources of error is the supposition or the taking for granted that others will see and comprehend human nature as we see and comprehend it. An individual judges a social problem from his or her understanding. He or she has longings, desires, emotions, and sensations, and he or she imagines that others have the same sensations, that they will respond to the same stimuli in exactly the same manner and with the same degree of intensity in a given circumstance, in a definite social order.

There are often greater differences between individuals of the same race than between individuals of different races. Some are more richly endowed with more highly evolved nervous systems. If we wish to understand the basis of a superior faculty, we study how the nervous system of the individual has

4

become specialized. In the same way if we wish to understand the inferiority of individuals we study in what way their nervous systems are defective. It is this differentiation of the nervous system which separates man from man more effectually than geographical isolation in our modern civilization. The period of reaction to tactual, to auditory, to visual sensations, depends upon the physiological condition of the central nervous system.

Animals possess eyes, the structure is apparently the same, but what a difference in function. Human beings possess hands which are apparently alike in structure, but what a vast difference in delicacy of touch, in muscular sensibility between them. Some have taste highly developed, as, for instance, wine tasters or tea tasters. With many very often the sense of taste is defective. Cutaneous sensibility, with some is developed to an abnormal degree. Persons with keen bodily perception are often affected by changes in the weather, or shiver at the approach of particular individuals, or feel approaching danger. Many animals have this faculty developed to a higher degree than human beings. Again, we have thick-skinned individuals with very slight cutaneous sensibility. This has passed into the popular expression that a person is thick-skinned, you can't hurt or affect him; or it is often said that he or she is callous, hardened, unfeeling, insensible to anything you may say or do.

We do certain things, because in the doing we derive satisfaction and pleasure, we avoid doing certain things because they give us pain. If we study the physiology of pleasure and pain, we find that the person with highly developed bodily perception and the thick-skinned individual are two widely different animals; and this difference arises from the fact that the same stimulus applied to the two individuals will vary in intensity and therefore will produce a different effect. In Michael Foster's " Text Book on Physiology," he says, that " a slight stimulus, such as
" gentle contact of the skin with some body, will pro-
" duce one kind of movement; and a strong stimulus,
" such as a sharp prick applied to the same spot of skin,
" will call forth quite a different movement. When a
" decapitated snake or newt is suspended and the skin
" of the tail lightly touched with the finger, the tail
" bends towards the finger; when the skin is pricked
" or burnt, the tail is turned away from the offending
" object. And so in many other instances.

" It must be remembered of course that a difference
" in the intensity of the stimulus entails a difference in
" the characters of the afferent impulses; gentle contact
" gives rise to what we call a sensation of touch, while
" a sharp prick gives rise to pain, consciousness being

6

"differently affected in the two cases because the "afferent impulses are different." The difference in the intensity of the same stimulus applied to the sensitive person, to the thick-skinned person, and the diseased person would affect their consciousness differently. The stimulus, which would be so intense to the sensitive person as to produce pain, would to the thick-skinned person only produce a sensation of touch. The coarse brutal word which would give rise to pain in the sensitive individual would give a sensation of satisfaction to the person with slight sensibility.

If pleasure and pain actuate our movements or determine the character of our movements, we have ample proof that influences which give pleasure to one class of individual may give pain to another. More light is thrown on this subject in another paragraph in Michael Foster's "Text Book on Physiology": "The clinical histories of diseases of the spinal cord in "man bring to light in a fairly clear manner a fact of "some importance, namely, that the several impulses "which form the basis of the several kinds of sensa-"tions, of touch, heat, cold, and pain, and of the "muscular sense, are transmitted along the cord in "different ways and presumably by different struc-"tures. For disease may impair one of these sensa-"tions and leave the others intact.

7

"Thus cases of spinal disease are recorded, in which on one side of the body or in one limb ordinary tactile sensations seemed to be little impaired, and yet sensations of pain were absent; when a needle was thrust into the skin no pain was felt, though the patient was aware that the needle had been pressed upon the skin at a particular spot; and conversely in other cases pain has been felt upon the insertion of a needle, though mere contact with or pressure on the skin could not be appreciated. Again, cases are recorded in which the skin was sensitive to touch or pain, but not to variations of temperature; it is further stated that cases have been met with in which cold could be appreciated but not heat, and *vice versâ*."

Many persons cannot be affected with kind words and mild treatment, but must be dealt with harshly or firmly. The psychical appreciation of the slighter stimulus is blocked, it is analogous to the one in whom touch could not be appreciated but on the insertion of the needle pain could be felt.

Certain poisons in the blood augment the excitability of the central nervous system, others deaden the sensibility. The fatigue products in the blood have a depressing influence on the central nervous system; imbecility, stupidity, dullness, imply lessened excitability of the nervous system. To arouse dull

8

or stupid people it requires a stronger stimulus than it requires for normal individuals, just as it requires a stronger stimulus to arouse a tired animal into action than a fresh one.

In the same way that we build insane asylums to house our insane because they have lost their mental balance, so we build pauper institutions for those who have lost their physical balance. The vagabond, the pauper, is as much born and made one as the man of insane temperament under stress demonstrates his neurotic heredity and the criminal his pathological condition. The terrible effects on posterity of depleting our workers and then allowing them in ignorance to breed is the burning question for humanity. The physical condition of the population of a manufacturing town is proverbial.

It is said that in a new country where the land has not yet been appropriated, there is no such thing as the unemployed or the pauper. When colonists first settle upon a piece of land there is plenty of outdoor exercise, manly pursuits, work which does not cause physical deterioration. But after a time as population increases and sedentary occupations take the place of active pursuits, crowded enclosed workrooms supplant work in the open air, the energy of the workers is gradually sapped by artificial life in cities, and they become the progenitors of a class physically enfeebled, spiritless, incapable of sustained effort. Work is carried on by means of the contractions of muscular fibres. Michael Foster says, in his "Text Book on Physiology," that "inter-"ference with the normal blood stream is followed "by a gradual diminution in the responses to stimuli "and the muscle loses all its irritability and becomes "rigid . . . . with regard to the quality of the "blood thus essential to the maintenance or res-"toration of irritability, our knowledge is definite "to one factor only, viz. the oxygen. If blood "deprived of its oxygen be sent through a muscle "removed from the body, irritability so far from "being maintained seems to have its disappearance "hastened. In fact, if venous blood continues to be "driven through a muscle, the irritability of the "muscle is lost even more rapidly than in the entire "absence of blood. It would seem that venous blood "is more injurious than none at all. If exhaustion be "not carried too far the muscle may however be "revived by a proper supply of oxygenated blood." The pallid faces and stunted growth of some of our town-bred workers tell their own tales. If exhaustion be carried too far in the living organism fatigue ensues, and in fatigue the muscles are slow to respond to stimuli. Individuals who are tired move and think more slowly and are less energetic. It can

10

be imagined how terrible are the physical results that ensue to those whose normal condition is one of fatigue.

Power of endurance in individuals is not equal; so that we could not say that eight hours' work or that less or more is beneficial to all alike. One may work eight hours continuously and not be exhausted, whereas another may be totally exhausted in six. Physicians warn us that if we do not allow sufficient rest to a tired organ to recuperate, waste products accumulate producing poisons which are a fruitful source of disease. The most active agent in generating the unfit is fatigue poison. If a large percentage of histories of family degeneration can be traced in the offspring of parents who have passed the prime of life, how much larger must the percentage of family degeneration be that is due to physical exhaustion from overwork or the lack of sufficient light and fresh air.

After great physical exhaustion the stomach is tired, it is often unable to digest heavy and coarse food, nature calls for something light, liquid of some kind, broth, tea, alcohol, and the like. And to this cause may be assigned the reason why the consumption of tea, alcohol, and opiates of different kinds is so largely on the increase. Animals who are over-tired or sick turn away from food. Anæmia is brought on by insufficient nutriment. Persons suffering from anæmia become apathetic, listless; the brain if no longer supplied with sufficient nutriment becomes torpid, the vital activity is lowered. Such persons have no energy to make an effort, they have no power of taking the initiative.

It is said that one in every five of the population of London does die or is destined to die in a hospital, the workhouse, or pauper lunatic asylum. *Pari passu* with this statistical statement the cry is growing louder for more public institutions to house the incapable, and it is urged that all stigma should be removed from them.

In visiting one of these institutions a short time since, in one of the wards I saw a little child moving about with the aid of a chair, its body being too big and heavy for its legs; in another ward a nurse, who was carrying a baby covered with scrofulous sores, asked me if I would adopt it. The baby had no one to claim it and they were only waiting to find someone who would take charge of it. There were cases of hip disease, some had been successfully operated upon. There was one with spina bifida. The doctor took great pride in showing me a child on whom he had just operated for hare-lip; my attention was drawn to the success he had had in delivering a mother of an idiotic baby. What is the destiny of these children? They require able-bodied

2 *

nurses from their birth, and able-bodied physicians to spend their valuable time over them. They are scarcely ever able to shift for themselves, they are a care all their lives, and at last swell the ranks of the one in five who die in the hospital, the workhouse, or pauper lunatic asylum.

The relationship between the abnormal palate and the brain is being recognized by all physicians who have made any study of the subject. They are consequently enabled to predict that in all probability the child with cleft palate will either be semi-idiotic, a criminal, or a lunatic, especially if subjected to the stress of poverty or adverse conditions, in any case will add to the burden already heavy laid on the community by the incapable. And the chances are they will be among the five who will die in the workhouse, hospital, or pauper lunatic asylum.

The following extract I copied from a paper :—

"A woman named Abigail Cochrane, who has just died at "Kilmalcolm at 84 years of age, was a pauper from the cradle "to the grave. She was born in Greenock in 1807, and was "imbecile from her earliest youth. It is estimated that she cost "the public purse between £2000 and £3000."

As in the case of Abigail Cochrane, each one of our human failures adds a considerable item to the burden, already large, put upon the healthy useful citizens. And if our present industrial workers are overtaxed, overburdened, and under the strain their health is undermined, what benefit will their progeny be to future generations? How are superior qualities to be transmitted to the offspring, if for generations the economic pressure has been so great as to deteriorate the physical constitution of their progenitors?

Physiology teaches us that conscious life is the result of the nature of the afferent or sensory impulses which reach the central nervous system, the physiological condition of the central grey matter, and the efferent impulses the central nervous system gives rise to which result in different movements. We know blue from red because they differentially stimulate the retina. Consciousness is differently affected because the afferent impulses are different. We recognize two different sounds because they differentially stimulate the auditory nerves. Consciousness is differently affected because the afferent impulses are different. We know already that the *character* of the afferent impulses varies with the intensity of the stimulus, and if the central nervous system is thrown into activity by the summation of afferent impulses reaching it, those who from overexertion or disease have their sensibility or excitability lessened, their nervous systems are fed in

14

less degree. Such persons would require stronger stimuli than normal healthy individuals to produce a given effect, for their central nervous systems either do not react at all to a given stimulus or else very feebly, with the results that they are dull or stupid. In one an afferent impulse may be so intense as to invoke a nerve storm, with another it may be too weak to have an effect.

Many are so deficient in sensibility that although afferent impulses may be started by the most beautiful pictures, sculpture, divine strains of music, noble and humane examples, in fact the most sublime combinations of nature and art, they will awaken no response, they will arouse no efferent processes of noble thoughts and actions. This accounts for the fact that certain persons only take pleasure in vulgar low resorts and the companionship of coarse people. They seek their affinities. The saying is, that a man is known by the company he keeps ; in other words, his nervous system is similarly developed.

If we study the nervous system of the pauper class, we find that instead of their nervous energy being economically expended, there is lavish, uneven and wasteful expenditure which is of no great benefit to the individual nor to society. They are organically deficient; they inherit defective, ill-regulated nervous systems, or their nervous systems become

15

badly adjusted through irregular habits, bad training, or diseases. They are incapable of sustained effort. They prefer jobs to regular work, spasmodic efforts to work for a few hours or days, and these efforts are followed by a reaction of utter inability to make further exertion. They can assign no reason why any sustained effort is wearisome to the last degree. These characteristics are symptomatic of retrogression, or they are the reappearance of a more primitive type.

There are savages who will work hard to collect material things, and then will debauch and idle away weeks and months until the pangs of hunger compel them to make another effort to work. In this we have the simplest condition of economic pressure. It is said that the special characteristic of the savage is that he has no thought for the morrow. He eats until he can eat no more, then goes hungry until he finds more food. These very characteristics we see exhibited among our own savages. I saw a poor man, who said he was hungry and had been given some bread and cheese, eat until his hunger was appeased and then throw the bread and cheese which remained into the street; he could not or did not realize that in a few hours he would be hungry again. I have frequently seen bread thrown away by such and lying in the street. To them bread had been given once, it would be given again, or they would go hungry until the pangs of hunger compelled them to make a further effort to procure more. It is a waste of words to say that these individuals are paupers because they have not been careful, thrifty, and temperate. We might lecture for hours to them on the advantages of industry, we might urge our plea with the fervour of a divine oracle, the afferent impulses we give rise to arouse no response in those torpid brains. For our plea to have an effect they must be given new nervous systems and healthy rich blood, in other words, they must not be bred. It is characteristic of those organically defective that it is the voluntary part of their nature which is most affected. They have not the *will* to make any exertion, they fall into the conditions which circumstances place them. With the offspring of parents suffering from fatigue or other poison, compulsory education may be enforced, but our efforts will not be repaid by healthy useful individuals unless they spring from a healthy source.

Political economists have said that the conscientious, the right-minded, will not marry until they are in a position to do so, and herein is the *crux* of the social problem. The more highly developed human beings yield less and less readily to the dictates of sexual passion alone. They judge and consider consequences.

17

They profit by the experiences of others and therefore avoid doing that which will bring sorrow to those whom they love. High motives deter the fit from marrying until they are in a position to do so. Among the better classes marriage is being deferred more and more, the standard of living is becoming higher among them, and more time is given to education, whereas the unfit who are not deterred by any qualms of conscience or apprehension of consequences go on multiplying. And as the more highly developed are not perpetuated, or if perpetuated it is in fewer numbers, the thoughtless, improvident, degenerate, and diseased, multiply upon us.

An educated man made the remark a short time ago, " The cause of so much misery among the poor "to-day is over-population, it is their reckless indul-"gence in large families. I am too poor to marry, I "can't afford to have a family, I wish I could, and yet "I am called upon to pay taxes to educate and help "to support others' paupers." Here is a man who was accustomed to a certain standard of living, and who therefore did not care to have offspring who would not have the same advantages as he had had, or to have a family who might become a burden on others. An example of the conscientious not marrying until he could afford it, a result which is most disastrous in its effects on the quality of the human race.

3

18

A man may possess a noble character and have a magnificent physique, but if he do not perpetuate these qualities they do not survive. A man may be diseased, stupid or reckless, but withal he marries and raises a large family: his qualities are perpetuated, but it is not the survival of the fittest. Many men break their health down by overwork, and the terrible strain is seen in the physical condition of their children. Many men have not over exerted themselves, and have had no scruples about living on the charity of their relations or friends, and hence their children do not suffer from the depleted physical condition of their fathers; but are these children the survival of fittest? Moral checks which would appeal to the superior intellectual mind, do not influence the unfit. In the majority of cases they have not a nervous system sufficiently developed to appreciate these motives.

A great many seem to think that interference with marriages of the unfit will only give greater opportunities to races, lower in the scale of development who are multiplying so fast, to overcome and conquer the more advanced races. We have an example of this in the rapid multiplication of the negroes in America, who at some not far distant day will outnumber and overrun the whites if the rapid increase be not checked. Eventually, if America is owned and governed by negroes, would it be the survival of the fittest? The outlook is as ominous in Europe.

Mr. Raines states in his census of the population of India, that the returns show an increase of thirty millions in the population in ten years, the total being 285,000,000. Add to this number 400,000,000, or probably more in China, and it looks as if these vast hordes may yet overrun and wipe out Western civilization. With this spectre looming up in the distance it is considered a dangerous policy to advocate any theory which would tend to limit the population of Western nations. The argument holds good if we wish simply to limit the numbers of the population of the fit, but has no application with regard to the marriages of the unfit. An American child brought up in China, if it had a defective nervous system, will demonstrate it in China; and a Chinese child brought up in Europe, if born of diseased parents, will demonstrate its hereditary condition here. We find often that physical causes, not numbers, determine whether races shall be conquerors or conquered. Stamina often gives the victory to a race. Generalship indicates superior development of the general.

But in any attempt to raise the standard of humanity, to aid evolution, we must take into con-

20

sideration that it is not the survival of the fittest, but the survival of the unfit by means of their rapid multiplication in societies as presently organized.

Any cause which determines the mating of individuals has a direct influence on the quality of the human race. All artificial social inducements for the mating of unsuitable individuals are instruments for the multiplication of the unfit. To prove how detrimental our present social life is to the human race, we have only to ask how many of the marriages which take place would be consummated if there were no social inducements, no fear of public opinion, no regard for the law, if there were no other inducement but the fact that he is male and she female, and that they are physiologically mated. How many who to-day are not mated would under such a state come together and propagate, and how many who to-day propagate from other inducements than love would no longer not do so? One great cause of the rapid multiplication of the unfit over the fit, is our false social system which places so many obstacles to prevent the coming together of our best men and women.

How many opportunities has a girl to find her physiological mate in her little set—even if she were free to choose? Sexual selection has very little scope in our conventional system. Take the many instances of women who marry for a home, very often the only

21

choice between that and starvation, and ask if there could be a greater perversion of the sexual instinct. I have heard it said, "What a good marriage Mr. —— "has made, he married a girl with fifty thousand "pounds, more or less; *she* is ugly and unattractive, "but what a windfall for him." A suitable marriage is often considered the one which will relieve the man from his debts or the marriage which will raise him or her up in the social or financial world. Money bags are highly valued in the marriage mart, and the, at present, artificial sign indicating that you, the vulgar, might not know the value of this piece of human flesh, so we tabulate him Lord or Prince.

Thousands of examples might here be given of the marriages *de convenance* of old men and young girls, and of young men and aged women which are so frequent nowadays. When these marriages are fruitful they too often produce idiots, murderers, or otherwise unfit. There are many social barriers which prevent the respectable poor from making physiologically suitable marriages. A respectable working woman said, "I work with a great many "men, but after business hours I do not dare go "about with any of them, for immediately all kinds of "reports would be circulated which would ruin my "reputation." The reckless or unfit not being deterred by any false restrictions go on multiplying.

22

Under our present industrial system there is a strong tendency against the survival of the fittest. If we take the life histories of two men, one honest the other dishonest, we shall find that nearly everything is in favour of the clever dishonest man in a plutocracy. Stanley Jevons gives us an example of how reckless speculation can be carried on at the expense of a credulous confiding public. "It now " becomes possible to create a fictitious supply of a " commodity, that is, to make people believe that a " supply exists which does not exist. The possessor " of a promissory note or warrant regards the docu- " ment as equivalent to the commodity named thereon. " It is only necessary then to print off, fill up, and " sign an additional number of such notes in order to " have a corresponding supply of commodity to sell. " It is true that the issue of promises involves their " fulfilment at a future day; but the future is " unknown, and the issuer may believe that before the " fulfilment is likely to be demanded the price of " the commodity will have fallen. Thus, if pig-iron " warrants could be issued in unlimited quantities " (irrespective of the stocks actually in the stores at " Glasgow), an unscrupulous band of speculators " might perhaps make large profits by selling great " quantities of iron for future delivery. *After* " *suddenly and excessively depressing the price of* "*pig-iron they might succeed in gradually buying up enough at lower prices to meet* the warrants when presented. This kind of 'bear' operations has certainly been successful in other markets."

"About ten years ago it became the practice to rig the market as regards the shares of particular joint-stock banking companies. A party would be formed, perhaps *owning none of the shares of the elected company*, and they would proceed to sell considerable quantities of the shares, hoping so to *damage* the *reputation* of the company and *lower* the *value of the stock* as to be able to buy up enough before delivery would be required. This noxious kind of speculation was checked by an Act of Parliament (30 Victoria, c. 29, 1867), which now requires the seller of bank shares to specify the numbers of the registered proprietors of the shares which he is selling for future delivery." In another paragraph of the same work, a further example of modern business transactions is given. The italics are my own in all three paragraphs. "Great injustice arises in some cases from this defective state of the gold currency. I have heard ot one case in which an *inexperienced* person, after receiving several hundred pounds in gold from a bullion dealer in the City of London, took them straight to the Bank of England for deposit. Most of the sovereigns were there found to be

"light, and a prodigious charge was made upon the
"unfortunate depositor. The dealer in bullion had
"evidently paid him the residuum of a mass of
"coins, from which he had picked the heavy ones.
"In a still worst case, lately reported to me, a man
"presented a post-office order at St. Martin's-le-
"Grand, and carried the sovereigns received to the
"stamp-office at Somerset House, where the coins
"were weighed, and some of them found to be
"deficient. Here a man was, so to say, defrauded
"between two Government offices."

Examples may be given to illustrate how the inexperienced are at the mercy of clever speculators. A man wishes to possess certain railroad shares, so that he can control a certain railroad; he offers to buy these shares, but the possessors, knowing they have good value for their money, refuse to sell; then the clever speculator goes to work to bring pressure to bear to force them to sell their shares, by creating a panic, or by spreading rumours to damage their credit; by this means, if they have many promissory notes or bills to meet, or money tied up, they may be obliged to realize on their shares, and the great financier has accomplished a *coup d'état*. An instance of the modern code of business ethics came under my notice a few weeks ago. A millionaire tried to negotiate a bill of exchange with a friend; fortunately, the friend had been made aware that the rate of exchange had fallen, otherwise this clever financier would have made perhaps a handsome profit out of his friend. Many examples might be given how the dishonest man, by trickery, bribery, furthers a scheme and amasses a fortune. And with this fortune he can pay the cleverest lawyers to defend him, if necessary, with his ill-gotten gains. The honest man in the time of distress pays his debts honourably, aids his friends, and takes advantage of no one, and, the consequence is, becomes poor. They both have children, the dishonest wealthy man can have the best professors, send his children to the best schools, his children will have better food and clothing, in time of sickness have the best medical attendance; they will have frequent changes from town to country, opportunities to travel and see the world; and having a favourable environment and not being subjected to the stress of poverty or conditions which would develop the latent bad they will acquire polish, become cultured, and with these superior advantages our *jeunesse dorée* will have their pick in the marriage market, where very often our fairest and tenderest flowers are knocked down to the highest bidder. And it is said, Would you take the reward of merit away? The honest man's sons, whose father has become impoverished, perhaps while they were still

26

young, may even have to do manual labour, they will be subjected to the stress of poverty, they will not have the best professors that *money* can procure in this world, where superior opportunities should be the reward of merit; they may have no opportunities to acquire polish and refinement, they may not become cultured in the terrible struggle with poverty.

Take an example of two men who are in the same line of business and are competing against each other. They are both individualists, and we are told that by their competing, prices are kept down and that we get a superior article for our money, and so forth. One of these men has capital to the amount of two hundred thousand pounds outside of his business, and the other has capital to the amount of one hundred thousand pounds. The one who has two hundred thousand pounds says, I can afford to undersell or underbid or carry on my business at a loss of one hundred thousand pounds; by that time I have ruined him, I have attracted all his customers from his establishment to mine, and when I have the monopoly I can gradually raise my prices till they give me a fair profit and my business is doubled.

This course is being pursued in nearly every line of business, the margin of profit is being run so close that old-established firms are obliged to sell out or continue business at a loss. I know of many instances where men have said, I can't take your order at a certain price for it will leave me no profit, I have to pay my workpeople, my rent, and support my family, and I can't do that unless I make a fair profit. Or, again, it is said, I can't do the job for such a price, but in many cases the man will be forced to become a sweater and will turn round and say to the workpeople, If you will work at so much an hour, I will take the job, I will make no profit out of it, if you don't feel disposed to work for so little, the order will be given to a competing firm who will perhaps do the job at a loss to attract a customer. I wished to investigate this subject for myself, and I have taken special pains to compare prices. For example, I went to shops which only sold certain articles, and then to universal providers or general providers. I found in one instance at a small shop the price was just double what I could buy the same article for at the large shop which sold nearly everything. I said to the proprietor, I can buy this at M——'s for fourpence. He took his book out to show me that he paid his manufacturer more than that wholesale. He said to me, they must have bought in a job lot or are selling them at a *loss* to attract customers to buy other things on which they make their profit.

In other instances, with other articles I had to

pay four shillings to the special dealer which allowed a *fair* profit, for what I could buy for three shillings within a penny or two more or less at the general providers. I found that the cheaper price was *cost* price to the manufacturers. Of course, the big houses are attracting the customers away from the small retailers and eventually they must go to the wall. As long as the small retailer was making a profit at *fair* prices, he was an employer, but with ruin staring him in the face he is obliged to give up his business and goes as an *employée*.

A hypothetical case will illustrate the effects of the modern tendency of concentrating several small industries into one large establishment : five small shops sell trimmings and buttons, the sixth, a big shop, sells these articles cheaper, or perhaps at first sells them at cost price to attract the customers, so that the people who require these articles go to the big shop. Therefore the five small shops are obliged to close, and the five former employers go to the big shop and become *employées* in the trimming department. This is said to be a departure in the direction of progress, because labour requires time—so time enters as an important factor in determining cost, and the time is saved in buying many articles at one shop, instead of being obliged to go about to many shops to find the required articles. The men, however, who have gone over to the big shop do not receive a share in the profits, but the sixth gets the former profits of the other five. The children of the five, the majority, fall several grades lower in the social scale, because with their altered fortunes they are obliged to leave school earlier, and the time they were giving to mental and physical culture must now be given up to work for bare subsistence. Whereas the children of the sixth, the minority, rises several grades higher.

I asked a saleswoman in one of these shops if the women received pretty fair wages. She answered, "We are the poorest paid workwomen in the City, " the firm makes such small profits they can't afford to " pay more." I asked her if they had any difficulty in getting workwomen. "No," she answered, " they " are never obliged to advertise, there are hundreds on " the list ready to take the place of one who falls out." I need scarcely say that many of the applicants were from the other ruined firms.

Before we can be quite sure that centralization of wealth and industries is in the direction of progress, the bodily degeneration caused in the production must be taken into account as part of the cost against the value of the utility. If labour must be regulated by supply and demand, the quantity of inferior people will create a demand for a quantity of

30

inferior goods. Our requirements demand certain economic goods ; in proportion as our taste becomes more highly educated the more difficult it will be to satisfy it. It is urged that the poor have so many useful things that formerly they were not able to possess, because they can be turned out by the quantity, and very cheap. The utility of certain articles of apparel is of more importance than the beauty. We get the hundred indifferently-made coats to-day where we used to get the one well-made before. And my opinion is, that we had better go back to sheep skins, because the utilities are used up or worn out, and have left no product, whereas the beautiful thing left the product in a developed æsthetic sense. Æsthetic taste enters very considerably into the value of an object. A picture may have no value to the vulgar, uneducated eye, but to the connoisseur who discerns its worth it has great value. The being able to appraise the value of an object, pre-supposes a faculty in the appraiser which the majority of ordinary people do not possess. So, then, if we wish a demand for this superior article, we must educate and develop this faculty in the individual.

If the great artist has no one to appreciate his genius it goes begging, if the superior workman gets no one to buy his work he falls back into the ranks of the many who supply the demand for the uneducated, undeveloped taste.

Among other causes which conduce to the rapid multiplication of the unfit, it has been suggested that the reason why the poor are so prolific is that they are underfed, that abundance of rich food lessens fertility. From experiments on flowering and fruit plants, it has been proved that by checking the *nutritive* conditions of the plant the reproductive power is increased; the roots of fruit trees are cut in order that they may yield abundant fruit. Darwin has already mentioned this subject in his work on animals and plants under domestication. He refers to some authors who have attempted to show that fertility increases and decreases in an inverse ratio to the amount of food. Darwin goes on to say, " This strange doc-
" trine has apparently arisen from individual animals
" when supplied with an inordinate quantity of food,
" and from plants of many kinds when grown on
" excessively rich soil, as on a dunghill, becoming
" sterile." In another chapter Darwin again refers to this subject in a paragraph, beginning, *sterility from the excessive development of the organs of Growth or Vegetation*. " To make European vegetables under
" the hot climate of India yield seed, it is necessary to
" check their growth; and when one-third grown,
" these are taken up and their stems and *tap-roots* are

"cut or mutilated. So it is with hybrids; for instance, "Professor Lecoq had three plants of Mirabilis which, "though they grew luxuriantly and flowered, were "quite sterile; *but after beating one with a stick until* "*a few branches alone were left, these at once yielded* "*good seed.*" There is other evidence bearing on this subject, especially the experiments of M. Maupas and others with infusoria, which have shown that when food became scarce the conjugal appetite increased, and when food was plentiful there was no conjugal inclination. There are certain worms which produce parthenogenetically when food is plentiful, and sexually when food is scarce, showing the intimate relation between nutrition and reproduction.

Opposed to this theory is the statement that population increases with the increase of food, that is, the number of animals increase rapidly where food is plentiful, and in times of dearth or scarcity the number decreases. The increase is ascribed to the increase in the number of marriages in prosperous periods. But among the very poor, where the increase is most rapid, I do not think this increase of population is due entirely to the *legal* marriage. It is very difficult to reconcile the naturalist's and the political economist's theories; the naturalist's that reproduction is augmented by the scarcity of food, the political economist's theory that population increases or diminishes inversely with the price of corn. Other checks may be at work which tend to obscure the value of the first theory; it may be that the food supply, falling below the minimum necessary to sustain life in a healthy condition, causes diseases or otherwise incapacitates human beings in the struggle against adverse conditions in an industrial crisis.

To sum up some of the principal causes in the rapid multiplication of the unfit, we may class them under two heads, namely, Physiological and Psychological.

Among the probable Psychological causes are:—

(1.) The more intelligent the individuals the more they think of consequences and the less likely are they to be influenced by sexual passion alone. Later marriages among the upper classes with the result of having fewer children, and if too long deferred the marriages are infertile. The improvident therefore would marry first and would rear the largest number of offspring. The sense of responsibility developes with age, but the very poor marry at very early ages.

(2.) Among the unfit easier modes of becoming acquainted, less prudery, more freedom in the intercourse of the sexes.

(3.) The mystery and secresy which envelopes these natural functions, too often create a morbid desire which often leads to masturbation and other practices.

(4.) Marriages among the upper classes for money and position, or the marriages of those who have not sufficient opportunities under our present social decrees to seek and find a more suitable partner.

(5.) The sexual passion excited by the intermingling of the sexes in overcrowded tenements; whole families often sleeping in one room. A lady who has a home for girls to help them through their first confinement, and to save first offenders, if possible, said: "It is appalling the number of "girls who come here who have been seduced by "their own brothers."

Among the probable Physiological causes are :—

(1.) Marriages of the immature, those who have passed the prime of life, or the physically exhausted, which produce offspring lacking in vigour and mental power, and only too often absolute idiocy is the result.

(2.) Inbreeding, especially if the parents are very similar, which intensifies morbid tendencies, the offspring from these marriages suffer from impaired mental power and lack of vigour; although close inbreeding gives a tendency towards idiocy, it also inclines towards insanity. It is said that insanity is one of the scourges of Newfoundland where intermarriage obtains. This also may be the result of the parents having been subjected to the same conditions of life.

(3.) Too great a difference between parents, for instance, cross-marriage, which give a tendency to reversion, as Darwin has so clearly demonstrated. Disease affecting the reproductive system also favours a tendency to reversion. It is said that cross-breeding is analogous to disease by producing an abnormal condition of ovum and sperm. From these marriages are supplied our criminals and the monstrosities. As Darwin says, "A similar tendency to "the recovery of long-lost characters holds good even "with the instincts of crossed animals. There are "some breeds of fowls which are called 'everlasting "layers,' because they have lost the instinct of incuba- "tion; and so rare is it for them to incubate that I "have seen notices published in works on poultry, "when hens of such breeds have taken to sit. Yet the "aboriginal species was of course a good incubator; "and with birds in a state of nature hardly any instinct "is so strong as this. Now, so many cases have been "recorded of the crossed offspring from two races, "neither of which are incubators, becoming first-rate "sitters, that the reappearance of this instinct must be "attributed to reversion from crossing." The reappearance of long-lost characters also occurs when disease affects the ovaries and testes. In disease of the ovaries, characters which have been latent in the female may become actual or effective. It seems incredible with our modern ideas of ethics that instincts or characters which it may have taken thousands of years of civilization to modify or suppress, should reappear as the result of influences which have affected the reproductive organs of the parents in a single generation. It especially strikes us with horror when we realize how common diseases of these organs are to-day.

(4.) Artificial preventive checks, which are more within the reach of the well-to-do classes than the very poor. Especially as these would affect the reproductive organs unfavourably and by this means gives a tendency to reversion.

(5.) The extreme susceptibility of the reproductive organs to changed or unnatural conditions, whether these be psychical or physical. The perversion of the sexual instinct often destroys all natural feeling, instance ancient and modern infanticide, fœticide, overlaying, suffocating infants, slow starvation, the frequent falls which are only too often premeditated, and many other instances of perverted natural feelings. The accounts of the perverted sexual instinct among certain tribes and even among modern nations may be due to unnatural conditions affecting the reproductive system; and to this fact also may be attributed prehistoric cannibalism, anthropophagy. Darwin remarks under the heading sterility from changed conditions, showing the extreme susceptibility of the reproductive organs to these changes: "When conception takes place under confinement, "the young are often born dead, or die soon, or are "ill-formed. This frequently occurs in the Zoo-"logical Gardens, and, according to Rengger, with "native animals confined in Paraguay. The mother's "milk often fails. We may also attribute to the dis-"turbance of the sexual functions the frequent "occurrence of that monstrous instinct which leads "the mother to devour her own offspring."

(7.) Disease, unless it directly affects the reproductive organs, seems to have no direct influence in lessening fertility. Diseased animals if left to nature would in all probability die off. Medical science, however, keeps them alive in order that they may propagate their kind.

(8.) The evident correlation between the brain and generative organs, the more the brain is exercised or when the female is given abundant rich food; in fact, the more the vegetative organs are developed in the female, especially where this is excessive, sterility is often the result. We have analagous cases in rich seedless fruit and double flowers.

The extreme delicacy of the females of the upper

classes from their artificial life is also a cause of lessened fertility. Also the sowing of the wild oats of the young men of the upper classes, is too often the cause of the sterility of the females whom they marry.

To disease in the parents may be attributed the largest share in generating the unfit. I read some time ago an article on Hydrocephalus, written by a doctor, who states that hydrocephalus occurs in about one in three thousand confinements, and that if syphilis, which is *such a common disease*, were the cause of hydrocephalus, why hydrocephalus would be more common still. To look up the family history of a patient is now a common practice. I had a girl in my employ whose conduct was very strange; I found on enquiry that her father and brother were in an insane asylum, and it will only be a short time when she will have to be placed under restraint.

The best minds of to-day have accepted the fact that if superior people are desired, they must be bred; and if imbeciles, criminals, paupers, and otherwise unfit are undesirable citizens they must not be bred.

The first principle of the breeder's art is to weed out the inferior animals to avoid conditions which give a tendency to reversion and then to bring together superior animals under the most favourable conditions. We can produce numerous modifications of structure by careful selection of different animals, and there is no reason why, if society were differently organized, that we should not be able to modify and improve the human species to the same extent. In order to do this we must make a religion of the procreative principle. Our girls and boys must be taught how sacred is the life-giving principle. The most wonderful of all the forces at work throughout nature.

Our young men and women should realize the purpose for which they are uniting in the holiest bond of physical life. And by this means we would have inaugurated the upper million and the lower ten. Any social conditions which tend to transpose these terms are subversive of the true interests of humanity.

VICTORIA C. WOODHULL MARTIN.

CHAPTER 8

The Scientific Propagation of the Human Race; or, Humanitarian Aspects of Finance and Marriage. The Science of Well Being. A Lecture

Introduced by Michael W. Perry

> *But man is not content, through the advantages conferred on him by reason of his intellectual superiority, to debar natural selection from acting freely. He builds hospitals, asylums, and poor-houses; and medical experts do all they can to keep alive the unfit brought together in these institutions, and destined, should they survive, to perpetuate a deteriorated race. It is a mistake, moreover, to imagine that natural selection always acts in the direction of progress. A negro survives in the interior of Africa where the European succumbs: is the negro, therefore, the fittest to survive? An unfavourable environment may foster the undesirable individual, whereas it would kill the ideally fittest.*

As Victoria Woodhull explained on the title page, this booklet comes from a lecture she delivered in New York City in November of 1893, which was itself based on speeches she gave "throughout America from 1870 to 1876." As a result, this booklet gives a unique glimpse into her early attempts to promote eugenics in the United States. It also explains, more clearly than any of the others, what Woodhull believed. Here we have her arguments honed by constant repetition and sharpened by questions from audiences.

The quote above (page 10) is an almost perfect statement of the world as seen by eugenists. The disabled ("asylums"), the chronically sick ("hospitals"), the poor ("poor-houses"), and some races ("the negro") were held in contempt. All were "undesirable" and all, if allowed to reproduce, were likely "to perpetuate a deteriorated race." Keep in mind that, when Woodhull was speaking on this topic in the mid-1870s, slavery had been eliminated only a decade before.

In the decades to follow, similar arguments would come from others, as I describe in *The Pivot of Civilization in Historical Perspective*. For those who thought in Darwinian terms, the enormous improvements in public health during the second half of the nineteenth century had a depressing downside. The life-preserving abilities of the human race in general and the kindness of some was sending us spiraling downward into a "deteriorated race." Woodhull, however, was wrong about one thing. Most of the decline in death rates did not come from institutional care. It came from public health measures such as clean water supplies, food inspection, and less crowded housing. By controlling infectious diseases, those changes benefited everyone, rich and poor.

Woodhull recognized that the problem was not with the theory of evolution itself. Despite Darwin's claims at the end of *The Origins of Species*, evolution does not come with a promise that those who are superior in some fashionable sense will win the reproductive battle and become the humanity of the future. The "fittest" are merely those who have the most offspring in a particular environment. In the African tropics, Europeans did poorly in comparison to the native population. In industrialized societies, affluent people had fewer children than the factory workers they looked down on. What Woodhull and those like her considered the "ideally fittest" did not automatically win the birth race. In that uncomfortable fact lay the roots of movements that would in the next century champion sexually segregated institutionalization, sterilization, birth control, immigration restriction, and finally legalized abortion.

Unlike *laissez faire* Social Darwinians, Woodhull did not want to return to a past where nature 'red with tooth and claw' ruled, and the 'fit' lived while the 'unfit' died. Reflecting her roots in the political left, she wanted *more* government intervention rather than *less*.

> ...The modern scientific breeder accomplishes as much in a decade as nature would achieve in a thousand years. And so might the scientific adjustment of society accomplish in a quarter of a century as much toward uplifting and improving humanity as nature backed by *laissez faire* doctrinaires would perform in centuries. (19)

But how was that "scientific adjustment" to be carried out? Woodhull had a scheme that targeted both heredity and environment.

> ...Idealise sexual selection and improve the marriage system, and the individuals who are the products of defective breeding disappear; improve the economic conditions, and individuals who are the product of defective environments decrease. (13)

Idealizing sexual selection was the new eugenic religion that feminists such as Woodhull, Ellen Key, and Charlotte Perkins Gilman found so appealing. (Margaret Sanger watered the idea down to a vague mystique about sex and motherhood.) Think of it as a parody of Roman Catholicism's vows of celibacy. In Catholicism, the few who are deeply spiritual forgo parenthood to better serve God. In eugenic religion, the many who are inferior forgo parenthood and toil at menial jobs their entire lives, never marrying or having children, in order to create a new humanity that would regard them with utter contempt. Needless to say, it had no chance of succeeding outside a narrow, snobbish circle.

Improving marriage meant changing everything that results in less-than-ideal births. Woodhull mentioned one aspect when she wrote: "There is no greater check to progress than that woman should be obliged to procure her livelihood by trading on her sex, whether by way of marriage for mercenary considerations, or by yet more degrading expedients." (32–33) Those who marry young or women forced by family finances into marrying a feeble but rich elderly man are examples of what she meant. But are young women really going to turn a cold shoulder to their high school sweethearts or is a mercenary woman going to pass up the chance of instant wealth for a eugenic abstraction? Not if she's given any choice in the matter. Woodhull herself liked to marry rich.

In the end, Woodhull conceded that the "scientific adjustment" she wanted must look like the breeding of animals. "Unworthy individuals" must be "excluded from breeding" by force. (34) She called for three techniques.

First, men and women are to be educated "to the responsibility of becoming parents." (35) Her soft words hide a harsh reality. Quite a few people must be forced to concede that they would be bad parents and must remain childless. That's unlikely to happen, particularly since the less responsible someone is about their own children, the less likely they are to feel responsible for humanity in the abstract. Mere education would become coercion.

Her second scheme is more plausible. Public opinion must be taught "the importance of intelligent breeding, until there is a reaction in public sentiment against the crime of perpetuating infirmities." (35) The majority is to be persuaded that it is in their interest to keep others childless. Since people are more amendable to measures imposed on others, that scheme at least had a chance of working. But it also suggests that those who dissent must be silenced and that means an end to freedom of press, speech or religion, a least concerning sex, marriage and family. Think of hate codes applied to sexual behavior.

Even when Woodhull seemed to offer women "freedom of choice," it came with a nasty legal bite. The paragraph below follows one in which she criticized situations in which a woman was "legally married to a drunkard, an epileptic, or otherwise unfit individual." (39) According to her, such "a woman perpetuates a crime if she continues to have children by such a husband." To avoid becoming a criminal, Woodhull expects that woman to divorce a man that, at least in the case of an epileptic, might be a congenial husband. In place of her husband, Woodhull offers an unlikely society where such women and their children aren't faced with poverty. She also slides quickly past the fact that a eugenic rationale for divorce would offer men an excuse to exchange an emotionally troubled wife for a less complicated and prettier one. (Bolding added.)

> It is often said that if the legal ceremony as the standard of morality be abolished, every man who is tired of his wife will leave her for another, or when she becomes old she will be replaced by some younger and prettier woman. But my view is, that if the financial dependence of woman on man be abolished, those men who could not exercise self-control or who were not actuated by high ideals would be weeded out by the operation of sexual selection. **Give freedom of choice to woman by making her procreative function independent of, not subservient to, her daily wants, and then will be bred a better race of men.** (39)

The third scheme is the most direct. Public attitudes created by the second will justify social policies that forcibly prevents births using, "the absolute isolation from society of irresponsible or unfit individuals." (35) Call them state institutions, prisons, or concentration camps, it doesn't matter. These people are to be kept from contact with the opposite sex for most of their adult lives, as well as from society, so ordinary people won't feel compassion for their plight.

But isolating people in institutions for thirty to fifty years is so expensive, it can only be used for a small minority. Any scheme to dramatically improve humanity must deal with the Fifty Percent Problem. Set any sort of standard, and half the population will be below average, delaying for many generations any scheme to improve the average. Most of that fifty percent can't be persuaded that having children is irresponsible, because their children will do quite well. Nor is it likely that public could be persuaded that these people are committing "the crime of perpetuating infirmities," because it is obvious that most of them benefit society, since not everyone needs to be a rocket scientist. Finally, the real harm to society comes not from those who are slightly below average in intelligence, but from the talented evil, a group eugenists have never shown an interest in purging from the gene pool.

In fairness it should be pointed out that Woodhull placed herself among those who had given birth to an inferior child. The reason wasn't genetic, but rested on her belief that, "The marriages of the immature also curse humanity by producing individuals who exhibit every stage of mental defect." (14) There seems little doubt that this describes Woodhull's own experience.

> The case may be cited of a girl who was left in ignorance of the simplest laws regarding her being. She was a victim of this ignorance; she became a mother at fourteen. Her child is an imbecile. The father and mother of the girl love this grandchild and deplore the fact that it is an imbecile. They are too ignorant to realise what caused this human failure. (14)

Nor should we regard Woodhull as inherently cruel. She cared for that retarded son for the rest of his life.

Notice that Woodhull opened the speech beginning on the next page by referring to "agitation extending over thirty-seven years." Dating back from 1893, she was claiming that the debate over "scientific propagation" began about 1856, nine years before Francis Galton's 1865 magazine article and three years before Darwin's *The Origin of Species*. A brief search came up with one possibility, a 72-page pamphlet by a J. Soule that the Library of Congress dates to 1856 Cincinnati: *Science of Reproduction and Reproductive Control: The necessity of some abstaining from having children —the duty of all to limit their families according ... the different modes of preventing conception.* When that pamphlet came out, Woodhull was also in Ohio, and the nineteen-year-old mother of a young child she later called an "imbecile."

A PAGE OF AMERICAN HISTORY.

The Scientific Propagation of the Human Race;
Or, Humanitarian Aspects of Finance and Marriage.
The Science of well being.

A LECTURE
Delivered at Carnegie Music Hall, New York City, November 20th, 1893, and throughout America, from 1870 to 1876
BY
VICTORIA C. WOODHULL.

"The dreams of yesterday are the realities of to-day."—CARLYLE.

"I dreamed that stone by stone I reared a sacred fane, a temple, neither pagoda, mosque, nor church, but loftier, simpler, always open-doored to every breath from heaven, and Truth and Peace and Love and Justice came and dwelt therein."—TENNYSON.

It may safely be asserted that, after an agitation extending over thirty-seven years the questions comprehended in the term Scientific Propagation are now fairly launched, and they will be discussed until the vital importance of applying scientific principles to the breeding of the genus *homo*, as well as to the genera of other animals, be universally recognised.

No subject can be of such vital importance as the one which enables us to understand our birth and being. With the great advances of science of the last half century, we are gradually completing the history of the various races which inhabit and have peopled this globe. From a simple cell to the complete being, the origin, development, and progress of man are gradually becoming revealed. The difficulties which bigotry and ignorance have put in the way of a proper understanding of man are being overcome, and the mysterious awe which enveloped the subject is being dissipated by scientific research. We no longer look upon man as something apart in the universe, as a being specially created and consecrated. We recognise our animal origin; we dissect the various organs of the body; we study the functions of these organs; we dilate upon the passions and desires to which they give rise; we speculate upon the ultimate use to which they may be put. As we acquire a clearer and more definite knowledge of our animal origin and a knowledge of organisms lower in the organic scale, the question suggests itself, In what are we similar to and in what do we differ from these lower beings? We have organs analogous to those of other animals; we exhibit fear, rage, aversion, love, curiosity, wonder, sympathy, fidelity, pain, pleasure, æsthetic feeling, as do they. In fact, our passions, desires, instincts, have their foundation in, and are dependent upon, our animal organs. When an animal can put to various use a single organ, it is said to have advanced in organisation; when it loses the use of any organ which it formerly possessed, it is said to have retrograded, degenerated. It is the use to which man has adapted his various organs which gives him such a high position in the organic scale. His various senses have become modified, specialised in a thousand different ways.

But does this modification of structure give him the right to be called other than animal? Passions, desires, instincts are dependent upon animal organs;

but, it may be asked, is there not evidence that man does exist apart from his organs, although imprisoned in this house of flesh ? is not the *ego* independent of nerves which vibrate, of muscles which contract, of the blood which circulates, carrying its living food ? Is man's nature not his blood ? " No, certainly not ; but men's blood is their nature. Theophrastus, in his treatise on Ethics, discusses whether a man's character can be changed by disease, and whether virtue depends upon bodily health ; and to-day we know that it is so, but we do not act upon our knowledge. It is the malnutrition of the various tissues of the body that causes deterioration of the individual, acquired and hereditary. The nature is controlled by the blood, not the blood by the nature. The man suffering from heart disease, the melancholic, the hysteric, the optimist, each in his turn betrays to the physician the organ that is diseased. From the innermost fibres of the central cord and brain, to the skin outside and the chemicals circulating within the blood, influences which affect one and all, have their corresponding psychical effect. A healthy soul depends upon a healthy mind in a healthy body. Pleasures and pains are derived from the satisfaction of man's organic needs, as is the case with other animals.

Evolution of the animal into the man—how much does this convey to the thinker ? Man was not, then, always human : when did he first deserve the title of human being ? He was an animal like other animals ; but there came a time when man commenced to know good and evil, when he could *control* his animal instincts in accordance with his perception of good and evil. It is when self-control is developed that a man or woman will not yield to animal inclinations, at the expense of bringing misery to others. The human is developed when a man or woman can subordinate his or her desires to a higher good. I will not yield to temptation " may be said to be the first answer given by the human being to the animal being.

It is a well-known physiological law that the animal man reasserts himself very powerfully when the human has ceased to exert a restraining influence. Man, civilised man, humanised man, starts with the same organic passions as other animals, but in him they have become refined, elevated, exalted. He is capable of experiencing pleasures and pains of the intellect which animals cannot experience. This is a distinguishing feature of a human being, to be able to take pleasure in those things which appeal entirely to the intellect. In a low order of beings only those things excite pleasure which satisfy animal needs. The highly organised, sensitive human being who is at the top of the organic scale appreciates and derives pleasure from the cultivation of the beautiful—from poetry, music, painting, architecture, sculpture, and other humanising pursuits. In progressive higher forms of human beings the pursuit of happiness will be still more idealised. Degraded individuals see beauty in and derive pleasure from those things which are degraded ; elevated human beings take pleasure in those things which elevate. Each thing will be appreciated according to the standard of the individual. The ideals of religion have exerted a powerful influence in developing the human man, but there were other forces at work bringing about progress in the organisation of animal man anterior to religion.

9

Darwin, in his great theory of natural selection, offered an explanation of how progress in evolution was attained. He endeavoured to show why lower forms of animal life give way to the higher. Natural selection is based on the survival of the fittest. Nature selects in each condition those which are fittest to survive. As the environment changed it produced corresponding physiological changes. The animals who were not thus modified would not survive.

With the acceptance of this theory, many have jumped to the conclusion that progress is the natural order of things, that everything must needs move in the direction of progress. Hence it is believed that only the fittest survive. Now Darwin himself, quoting Wallace, argues that man, after he had partially acquired those intellectual and moral faculties which distinguish him from the lower animals, would have been but little liable to bodily modifications through natural selection or any other means. For man is enabled through his mental faculties to keep, with an unchanged body, in harmony with the changing universe. He has great power of adapting his habits to new conditions of life. He invents weapons, tools, and various stratagems to procure food and to defend himself.

The lower animals, on the other hand, must have their bodily structure modified in order to survive under greatly changed conditions. They must be rendered stronger, or acquire more effective teeth or claws, for defence against new enemies; or they must be reduced in size so as to escape detection and danger.

But man is not content, through the advantages conferred on him by reason of his intellectual superiority, to debar natural selection from acting freely. He builds hospitals, asylums, and poor-houses; and medical experts do all they can to keep alive the unfit brought together in these institutions, and destined, should they survive, to perpetuate a deteriorated race. It is a mistake, moreover, to imagine that natural selection always acts in the direction of progress. A negro survives in the interior of Africa where the European succumbs : is the negro, therefore, the fittest to survive ? An unfavourable environment may foster the undesirable individual, whereas it would kill the ideally fittest.

We are very careful about the soil and the aspect in which we put a choice tree or plant ; but common or hardy trees and plants will grow and multiply without skill and care. Are they, however, the fittest to survive ? Whether certain individuals will survive and others succumb in certain conditions will depend upon the power of resistance to or affiliation with the particular surroundings which those individuals possess. The man of coarse fibre and low mental capacity will survive in low and debasing conditions which would kill the highly developed sensitive man. The former, therefore, surviving and multiplying, would not depend upon his fitness in the direction of progress.

To understand this fact a little better, it is only necessary to ask, What determines the environment in our modern industrial system ? Money—the purchasing power. Of a dozen men who are working at a trade, one will evince superior talent and receive higher wages than the other eleven, because superiority should be rewarded.

In the second generation this one superior man who has enabled his family to have superior advantages

II

will be represented by four or five children, whereas the other eleven who have had the disadvantages of an inferior environment will be represented by fifty or more. Moreover, in a polity ruled by a majority vote, the latter will determine who shall and who shall not be put into office. In a social system which gives superior advantages to the few who acquire the wealth, how does humanity benefit as a whole if those individuals do not perpetuate themselves? Will the standard of humanity be raised into progressively higher and higher forms? If wages are low and money scarce, the majority will be forced to eat inferior food, wear insufficient clothing, to live in unsanitary dwellings or shops, or to work at trades prejudicial to health. If the persons who are overworked and underfed are those who are continuing their kind, the quality of the race must necessarily deteriorate.

Every individual is the product of two factors—breeding and environment. If a man be an idiot or a criminal, we have only to study how much is due to breeding and how much is due to environment, to know the why and wherefore of his being so. If a man be bred with ungovernable desires and propensities, the less stress is required in the struggle for existence to bring out or develop in him the latent tendencies which betray the criminal. The stronger the influence of the one factor, the less of the other factor is needed to produce the result. The amount of disturbance that is needed to upset any orderly arrangement depends entirely on the stability of the arrangement. If a man be bred well, so that in every way he is physically and mentally perfect, a more powerful stress in the struggle for existence would be required to make him do an irrational act. For instance, a man may be born with every physical advantage which even scientific propagation would bestow, but the stress from conditions of the environment will at last make him succumb, if it be continued long enough and be sufficiently severe. There are stresses which will upset the most stably constituted nervous system. The man willing to work but unable to find employment, seeing his wife and child starving, might be driven to any rash act; whereas in a favourable environment, without stress, he would become, probably, one of the noblest and most useful members of society. The greater his sympathy, the more unselfish his love, the greater his capacity for appreciating the might be, the more desperate will he become. Whenever you read of a crime in a newspaper, you can ask yourself, Is that individual a victim of bad breeding, or one of the martyrs of our ignorant social mal-adjustments? If a woman is a devil or a saint, she is so because she was bred under certain conditions, and because the influences of the environment which make the production of devils or saints acted upon her. The stronger and more intense the passions and the less developed the higher control centres, the slighter the provocation or stress that will drive the individual to rash acts or immoral conduct. On the other hand, no matter how well developed the control centres, if only sufficient stress be brought to bear, if anxiety and worry only continue long enough, or if a poison induced by disease continue to act on these centres, the individual will at last succumb. The more of the hereditary disorganisation present, the less stress will produce a given result. Disorderly conduct, bad actions, unduly displayed appetites, propensities, or desires, are simply a derangement of the higher nerve centres. If nervous strain, if economic pressure, does not upset the majority of average individuals, but is sufficient to upset some, we know then that these individuals have inherited defective nervous organisations; they are not bred up to the usual or normal standard. Breeding and environment, therefore, are the two factors which determine whether a man will attain a high standard of physical or mental perfection, or whether he will exhibit evidences of physical or mental degradation. The weak-minded, the drunkards, the paupers, the criminals, the imbeciles by nature are the product of breeding; the defective individuals by nurture are the product of their environment. Remedies consist in altering the conditions which give rise to the two classes. Idealise sexual selection and improve the marriage system, and the individuals who are the product of defective breeding disappear; improve the economic conditions, and individuals who are the product of defective environment decrease. In so far as reforms have the effect of raising the standard of material well being, they will diminish or eliminate crime, drunkenness, imbecility, pauperism, insanity, and poverty. In so far as reforms have the effect of ostracising improper marriages and encouraging worthy ones, defective individuals of the first class will be exterminated.

Apart from the fact that comparatively the few control the wealth, and are therefore able to command the superior environment, there are other forces at work determining the survival of the unfit. The underfed are more fertile than the overfed. It is significant that plants grown on poor soil run to seed. Cultivation has the effect of modifying the reproductive system.

Not only have the rich the superior environment which would insure the survival of the fittest, but the very power which money gives them of commanding rich and luxurious food contributes to sterility among them.

The overworked and underfed portion of the community have the largest families. Statistics show that among rich leisured classes families are comparatively small, whereas among the poor families are large. In poor, overcrowded dwellings the relations between the sexes are less conventional, and the stress of poverty favours early marriages, thus insuring the survival of those who from an ideal point of view are unfit. The prudent or more highly developed do not marry at early ages, and if their marriages are too long deferred, they are sterile.

The marriages of the immature also curse humanity by producing individuals who exhibit every stage of mental defect, from absolute idiocy to those who are simply stupid. The case may be cited of a girl who was left in ignorance of the simplest laws regarding her being. She was a victim to this ignorance; she became a mother at fourteen. Her child is an imbecile. The father and mother of the girl love this grandchild and deplore the fact that it is an imbecile. They are too ignorant to realise what caused this human failure. Marriage among the wealthier and better educated of the community is tending to be deferred. Among the poorer and most destitute the age of marriage has a tendency to become earlier. From these immature marriages offspring will be born who again will do little to raise the standard of humanity, who will not be deterred from procreating their kind from any considerations of right. Children

15

born during the period of maximum vigour of the parents have, as a rule, the greatest physical and mental vigour. What must be the result then in children born of devitalised parents? In the marriages of those individuals who have no natural affinity for each other, where other inducements than love determine their union, the offspring exhibit analogous physical defects. The case is the same in the offspring of parents who are too similar in defects of constitution.

Disease in the parents may affect the children, so that the result may be imbecility, insanity, or criminality, according to the manner and extent that the transmitted diseased blood acts upon the brain of these individuals. It is problematical whether the result of these criminal marriages will be an imbecile or a criminal. When the higher restraining centres which exert an inhibitory influence in the various organs cease to act, or are not developed, there is an abnormal activity or an unbalanced functional activity which may lead to immoral or criminal conduct.

When individuals are overworked or insufficiently fed, their brains will not assimilate knowledge as when their mental powers are fresh and vigorous. A celebrated physician wished to ascertain whether this was actually the case; he experimented upon himself by first tiring his muscles before some examinations, and he found that his mental energy was materially affected. In the North American Review a writer commented on the superior brightness and capacity of some of the rich men's sons at a certain college, as compared with those who were stuyding to become teachers. The latter were obliged to give lessons in order to pay for their school fees and lodgings, etc. Is it any wonder that their brains were not so fresh? Overwork and stress are bound to tell if they are only continued long enough. Even the strongest brain will succumb if it be overworked sufficiently. In a description of a play which was given in the poorest quarter of the East End of London it was stated that at some of the exciting scenes the management was obliged to stop the play until the poor, overworked, shattered nervous systems of the audience had time to recover their self-control. I have noticed at scientific lectures given to poor working men that many went to sleep. Rest was what they needed, not more work in mental exertion.

It has been said that the energy exhibited by a gang of navvies is not theirs, but that of their victuals; they simply direct it. The overworked and underfed exhibit the physical and mental defects consequent upon exhaustion of energy. They have not such vivid impressions associated with acts which appeal more to the reason. A horse when tired out with dragging a heavy load does not mind the whip; after a few days' rest in the stable, after he has stored up some potential energy, the whip will make him jump and fret and fume. If, from poverty, individuals have their blood impoverished or diseased, the character of the nutriment supplied to the higher control centres which bathes and nourishes them is altered. And as this alteration of the blood becomes more or less marked, its results are seen in the altered and often irrational conduct of individuals. This blood, by being deteriorated in quality, no longer supplies sufficient or efficient nourishment; it may be of such a character that it excites the nervous system, and rash or criminal con-

17

duct is stimulated ; or of such a character individuals become sluggish and indisposed to make any exertion, or are dull and stupid. When the higher nerve centres give way, the lower over-act. A man who in his normal condition would not do an improper act, when under the influence of liquor or some poison acting on these centres may become indecent, use foul language, and turn on his best friend. It is a fact well known to physicians that, under certain conditions of nervous hysteria, delicate and cultured girls will break out in language which it would have been thought impossible that they could ever have heard, much less have acquired. When we see how much depends upon the development and health of these higher centres, we realise how necessary it is to do away with such stresses as will attack or undermine them and drive persons to drink, to insanity, or into pauper institutions.

That power which modifies, checks, regulates our animal passions, develops the spirit of God within us, is the Soul. The Lord Buddha comprehended this great truth and founded his religion on the power of " I will." The higher control centres, as they become superimposed one on the other in the hierarchy of the nervous system, illustrate the mental evolution of the animal man into the human. From youth to manhood there is a slow development of the higher control centres. The child exhibits lack of self-control. Why ? Because these higher centres are not yet developed. Without the higher control centres man is simply a brute ; his lower appetites reign without any check.

The tendency of individuals is to satisfy animal

18

instincts, but when the higher controlling centres are well developed, these desires are restrained by the Spirit of God within His Temple, which says that the consequences will be thus and thus ; hence there is a direct check upon particular actions. Every influence which develops nerve centres controlling function develops consciousness ; as man and woman have their higher nerve centres more developed than other animals, their actions are more determined by conscious choice. When they have attained this mastery a conscience is developed. The passions of the civilised man may be as strong as those of the savage, but his power to control and regulate them is infinitely superior. The reason why the lower propensities are abnormally developed in idiots, savages, the insane, or other individuals, is that the brain never has exercised or has ceased to exercise a restraining influence.

Ancient philosophers recognised this great truth. They saw the struggle going on in man to gain self-mastery over his animal nature. Classical artists pictured man, half man, half beast, to symbolise the struggle. Where man had conquered the lower appetites and propensities he was human man ; where the struggle was still going on he was half man, half beast. We to-day can see the museum at Olympia, chiselled by Greek sculptors, " a Centaur about to carry off a woman, whom he holds with his left hand and right forefoot, while she in her struggles seizes him by the hair and beard. With his right hand the Centaur defends himself against Peirithous (human man), who advances to the rescue with his battle-axe raised." The best preserved group is the woman who has sunk on her knees, while the rearing Centaur clutches her hair with his left hand, and holds her fast with a hoof on her breast. The HUMAN part of the Centaur is wanting."

To portray the struggle between animal man and human man, giving the victory to the human, was the first inspiration of God in man. Thus man was inspired to realise that some day in the dim future the animal would be completely human, having mastery over himself and over nature's forces. To have conceived this ideal was the inspiration of God, equally with that which conceived an ideal heaven and hell.

Progress in evolution is accomplished by the elimination of the unfit; but how can the unfit be eliminated until it be ascertained who are the fit? When we have some conception of the ideally fittest, of the ideal man and woman who are influenced by ennobling beliefs, high aspirations, and godlike motives, we may then ask whether these are the individuals who are surviving and propagating their kind. If we find that this is not the case, we may know we are not moving in the direction of human progress. The modern scientific breeder accomplishes as much in a decade as nature would achieve in a thousand years. And so might the scientific adjustment of society accomplish in a quarter of a century as much toward uplifting and improving humanity as nature backed by *laissez faire* doctrinaires would perform in centuries.

What wonderful solicitude is shown in the breeding of choice animals, and what utter indifference in the breeding of boys and girls; whereas it ought to be the other way. Man is subject to the same laws; good food, pure air, contact action have as much influence upon man.

Political economists have said that the conscientious, the right-minded, will not marry until they are in a position to do so, and herein is the *crux* of the social problem. The more highly developed human beings yield less and less readily to the dictates of sexual passion alone. The human beings in whom the higher control centres are well developed will be able to consider consequences, and will not marry at the risk of entailing misery and degradation on their offspring. But still, if this high conception of honour prevents their marrying, these qualities which they possess are not perpetuated. On the other hand, those individuals who are not guided by reason, who are moved alone by animal instincts, will increase and multiply, and, consequently, those survive who are unfit from an ideal standpoint. High motives deter the fit from marrying until they are in a position to do so. Among the better classes, marriage is being deferred more and more; the standard of living is becoming higher among them, and more time is given to education; whereas the unfit, who are not deterred by any qualms of conscience or apprehension of consequences, go on multiplying. And as the more highly developed are not perpetuated, or if perpetuated it is in fewer numbers, the thoughtless, improvident, degenerate, and diseased multiply upon us.

A man may possess a noble character and have a magnificent physique, but if he do not perpetuate these qualities they do not survive. Another may be diseased, stupid, or reckless, but withal he marries and raises a large family. His qualities are perpetuated, but it is not the survival of the fittest. Many men break their health down by overwork, and the terrible strain is seen in the physical condition of their children. Many men have not over-exerted themselves, and have had no scruples about living on the charity of their relations or friends, and hence their children do not suffer from the depleted physical condition of their fathers; but are these children the survival of fittest? Moral checks which would appeal to the superior intellectual mind do not influence the unfit. In the majority of cases they have not a nervous system sufficiently developed to appreciate these motives. The improvident poor, finding so many agencies at work to relieve them from the consequences of their criminal folly, have the less inducement to deny themselves the gratification of their appetites, and conscientiously fill the hospitals with diseased men and women, and the asylums with cripples, scrofulous, syphilitic children, epileptics and idiots, at an annual outlay to the community of many millions in money, without one single remunerative item to show on the other side. It was traced out by painstaking research that from one woman who, like Topsy, merely growed without a pedigree, as a pauper in an almshouse on the Upper Hudson, about eighty-six years ago, there descended 673 children, grand-children, and great-grandchildren, of whom 200 were criminals of the dangerous class, 280 adult paupers, and 50 prostitutes, while 300 children of her lineage died prematurely. And it is estimated that the expense to the State of the descendants of this woman was over a million and a quarter dollars. Had the 300 children who died young lived to grow up, the loss to the community would have been still greater. Such loss does not represent merely the actual outlay, but the absence of the return which, if spent for legitimate purposes, might have been expected from it.

And how many such totally useless animal weeds

are the taxpayers, healthy and useful citizens, supporting throughout the country at the present time?

If the marriages of the unfit began and ended with themselves, there would be no necessity for the social scientist to evince any alarm; but when the fruits of these unions choke and sap the vitality of the fit, it is time the question of scientific propagation were discussed. It is essential to know why do some and why do not others have their higher control centres developed, because only those should be born who can reason whether an act is right or wrong.

There is a well-known biological law that the structure last formed is the first to disappear or degenerate. The longer a habit has become organic in the individual or the race, the more difficult it is to overcome. Animal passions and instincts which have been transmitted for thousands of years are firm and stable. Human instincts which are the latest to appear in the evolution of the race are unstable and disappear the first. Scratch the Russian and you find the Tartar; scratch the human and you find the brute.

Hughlings Jackson formulated the biological law that in disease, in the disintegration of the nervous system the last formed first disappeared; the human faculty, the will, the god in man disappeared, leaving the animal in full possession. In the maturing man and woman the last organ to attain complete development is the brain. When parents are overworked their offspring are sluggish, or exhibit feebleness of mind, and are the very ones who in their turn will marry without considering consequences. And if they do not become absolute parasites on the fit and strong, they will do little toward raising the standard of humanity. It is a very noteworthy fact that fatigue products retained in the blood produce in the offspring the same effects as those which they cause in the parent. In the individual they produce sluggishness and mental torpor, and if these overworked individuals marry, they reproduce offspring who exhibit imbecility or disinclination for exertion, which may go to the extent of producing absolute paupers or those who rise little above the level of the brutes. It is incredible that in this nineteenth century conditions should endure which will allow individuals to multiply in this depleted, devitalised condition. When we remember that individuals such as these will not be moved by high aspirations and will not be affected by noble examples, the apathy exhibited on this subject appears terrible, and all the more so when it is considered how much of the degradation of human beings might be avoided. Except, perhaps, the crime of allowing individuals to become devitalised, there is no greater crime than to permit such devitalised beings to continue their species.

Physical wrecks in their turn breed physical wrecks. Much has been said about the terrible effects upon offspring produced by drunken parents, and the relation between drunkenness and physical deterioration is worthy of consideration. When persons are overworked, fatigue products, which are poisons, accumulate in the blood. These poisons in the blood alter the circulation, which, in its turn, affects the nervous system. Now the craving for drink arises from the fact that beers, wines, and opiates raise the acidity of the blood, and thus counteract the effect of the fatigue poisons. The effect of these poisons upon the individual brain is the same in the offspring of these individuals. The one great cause of drunkenness is the overuse of the bodily functions. Though it may be said the chances are that the children of the drunkard will be inebriates, epileptics, idiots, or neurotics, the causes should be examined of which drunkenness is the effect. In this manner we may be able to check the increase of inebriates, idiots, neurotics, epileptics ; but the drunkards themselves and their offspring will not have their reasoning power or control centres sufficiently developed to appreciate the crime of perpetuating their diseased condition in their offspring. It is a terrible misfortune that those who are most degraded by their environment will be the least able to appreciate why they are so.

I have seen children, the results of immature marriages, exhibit every phase of mental defect from absolute idiocy to simple lack of vigour or weakness of mind. We have the opposite result when parents are in the prime of life ; then the offspring are better developed, more vigorous, larger in body or better organised in mind. In the offspring of immature parents there is not sufficient energy to carry the process of development on to perfection. Strange to say, while the offspring lose in mental power, endurance, vigour, or size, fertility increases with early marriages.

The defects which arise in the offspring when the parents are overworked or devitalised are similar in character to those which are due to immaturity. The offspring are defective in such a way as to indicate that the process of development had come to an end prematurely. If the full development of the organism depends upon the initial store of energy, we must study what fails to be developed as the store of energy gives out. The last part of the human organism to be developed is the brain ; if, therefore, the initial store of energy commence to fail before the organism has attained full development, the size of the body and the brain will be affected. In the stunted size and the undeveloped brain of the idiot we have evidence of the deficiency of the initial store of energy. Again, the energy may be sufficient to carry the organism a little further, and the individual will be an imbecile or simply weak-minded; or there may have been sufficient developmental energy to produce a perfect brain and the size of the body be unusually small: Pope was such an instance. There may be sufficient energy to produce a normal body and brain, but a lack of stamina and vigour which incapacitates the individual for any sustained effort. When there is insufficient energy, the defects commence to be apparent in the latest stages of development of individuals; and as the last qualities to be acquired are the higher nerve centres, the psycho-motor areas of the brain, these parts are most affected by the overworked and exhausted condition of the parents. If we consider paupers from the cradle to the grave, we know that they are not paupers by environment—that is, made so by such stresses as poverty or overwork—but that the initial store of energy was not sufficient to carry the organism to full perfection in the last stages. In paupers by environment, however, it is the external influences, such as insufficient food, overwork, worry, or diseases, which act upon the organism and set up a degenerative process in the individual. In old age there is a gradual enfeeblement of both body and mind. There is great similarity between very old persons and those who are devitalised while still young in years. Power of endurance is weakened, loss of memory becomes apparent, interest in surrounding events is diminished ; the same loss of initiative, the same decadence of body and mind. Natural death takes place as the forces of the body become exhausted. But from the incidence of stress or overwork this capital of potential energy may become exhausted in early life. Individuals do not all start with the same store of potential energy ; the capital of one is spent before that of another. Two individuals may start life with an equal store of energy, but the one is not subjected to stress, and his store of vital energy carries him on to a good old age ; the other has to contend with poverty and overwork, and his capital is used up prematurely, and he falls into a condition of non-usefulness.

An organism which receives a weaker initial impetus will differ from that which receives a stronger impetus in several ways. It will not pass through the early stages of its development quite so fast as the more viorous germ, but the difference in the rapidity of development will not be conspicuous until a comparatively advanced stage is reached, and then it will be seen that the one organism is continuing strongly and vigorously to develop, while the development of the other is coming to a standstill. It is the last stages in the development of the less vigorous germ which fail to be traversed. What are the last stages of development ? If we watch the germ through the process of its evolution into the adult organism, we find at first there is an indication of difference between the head and the body, then appears the spine—the foundation of the skeleton—then the heart, the intestinal canal, the lungs, the limbs, and so on. The several organs and systems grow, develop, and eventually become complete, reaching their final stage at very different periods of life, some attaining completion long before birth, others being incomplete until adult age is reached. So in the plant raised from seed : the first things to develop are the radicle and plumule, then the leaves, then the stem, and last of all the flower and fruit, which latter mark the attainment of adult age by the plant. Now the flower of the human organism is the highest portion of the nervous system. This is the portion in which development attains its supremest height. All the rest of the body is but, as it were, the foundation of and preparation for the highest nervous developments. The body is but a house for them to live in, an apparatus for them to act through, an organisation for them to control. They are the culmination and climax of the process of development. Hence, if development is not carried far enough, if it fails to reach the latest stages, if its forces are spent ere its full course is run, the part whose development will fail to be attained will be the highest nervous centres. And what is the evidence ? The evidence is that in close inbreeding not only do the late offspring of such inbreeding fail to reach the full size and stature of their race, but the highest nervous regions fail to attain the development normal to that class of organism. They are idiotic. The inbred fighting cocks stand to be cut up without making any resistance. The inbred pigs have not sense even to suck. All cases of idiocy, and of congenital imbecility, which is a lesser degree of idiocy, owe their defect to the weakness of the original impetus. The insufficient impetus given may be due to constitutional weakness or ill-health of the parents. In other cases, parents who are healthy and constitutionally strong in other respects may yet be unable to procreate healthy and vigorous children ; and if their children are stunted and feeble, it will be likely that they will be deficient in intellect. The failure in vigour that occurs with advancing life and weighs on the procreative function, with at least as heavy a stress as upon other functions, may be the cause of the lack of vigour in the impetus that is given to the germ.

The exhaustion of the nervous system from whatever cause deprives the individual of the power to master circumstances, to lift himself above the degrading influence of his environment, or to rise to any altitude of mental exaltation. After rest, change of environment, and cessation of worry, individuals have regained, to a certain extent, their former vigour ; but where the individual is too poor to take the necessary change and repose, the degenerative process follows a progressive course until the actual stage of non-usefulness is reached. Premature decadence of mental power, premature inability to make vigorous and active exertion, is nothing more nor less than the emptiness of the store-house of energy—an exhaustion of the gray matter of the brain. Hence, it is apparent that anything which causes an excessive depletion of this capital of potential energy will produce paupers. Whether an individual be rich, and is provided for by rich relations, or whether he be poor and has to be supported by the State, the manifestation of deficient vital power is alike in both. The first would not be called imbecile, but merely one who is not bright, while the poor man will be stigmatised a common pauper.

A young man full of vigour and hope goes forth to earn a living. He aspires to maintain his parents, as they have in the past maintained him, or he marries and endeavours to earn enough for himself and his wife and the little ones. All goes well at first, but hard times come. There are too many mouths to feed, or business is depressed. He makes heroic endeavours to find work, or overtaxes himself with unaccustomed work. There is not now, perhaps, enough food to go round, but still he has plenty of energy. But he is now living on his capital of potential energy instead of his income of daily strength, an income which would have taken him along in favourable conditions to a healthy and useful old age. But a few more months or years of strain and stress and he is a beggar. In the " Twenty-third Annual Report of Charities of the State of New York " there is a report on the removal of alien paupers, giving the condition of those persons who were found in the various poor-houses, almshouses, hospitals, and charitable institutions of this State. It says :

"It is gratifying to know that measures are, at last, in prospect of being matured during the present Congress for the better protection of the State from the evils resulting from the deportation to this country, through agencies in Europe, of so many pauper immigrants of the kind that have hitherto become dependent upon our public charities almost as soon as landed. The majority of this undesirable class this State has for years been burdened with the care and support of in almshouses, hospitals, asylums, and other institutions, for the reason that they arrive mainly through its principal port of entry, the City of New York. Our representatives in Congress should be requested to use their efforts to pass the Bill or Bills required to alleviate an evil of such magnitude in the unnatural and unjust burdens it imposes. Meanwhile, the usual appropriation of 5,000 dols. to enable this Board to carry out provisions of the law of 1880, for the removal and return to their homes abroad of such as find their way into this State by the way of Canada, or by any other channels, will be necessary as a wise and partially protective remedy, until federal action is assured and perfected in the direction of a more thorough one.

"The saving of future expenditure to the State by these modest annual ones is so enormous that it almost passes the belief of those not familiar with the subject. It amounts in economy to millions, as can be proved by estimating the cost of supporting in institutions 1,391 persons at only two dollars per week (making no estimate of the cost of the added 'plant' or buildings necessary to contain them), a total of nearly 2,800 dols. per week, and of 145,600 dols. per annum for the whole number.

"Estimating the duration of life of this class of dependents, had they been allowed to remain in the country, at the minimum average of fifteen years, the result of the wise forethought of the State in annually appropriating the small sums used since 1880 to return foreign paupers found in it—an ultimate saving of 2,184,000 dols., independent of the expense of housing them and providing salaried officers for their care-taking for that period of time, fifteen years—is definitely proved, and an illegitimate burden on our people thrown back where it originated, and where it justly and naturally belongs.

31

"Obviously no measure can be more prudent for the State than to protect itself, as far as it can, from the noisome sediment that forced pauper immigration from all Europe deposits, almost immediately upon arrival, in its institutions, supported by taxation, and also in those of its counties, cities, and towns, locally supported in like manner."

The question agitating the scientific sociologists of to-day is not whether the pauper who is destined to be a burden to the State be a Pole, an Italian, a German, Swede, or of any other nationality, but what produces the pauper who is liable to be a burden to any State. If it be breeding, then we must spread the necessary knowledge on this vital subject; if it be the environment, then we must do away with the economic condition which has as a result this human wreckage. They are either human failures or they are human wreckage.

The cost of the maintenance of the non-effective classes, whether they be few or many, is a futile expenditure of money and energy. If it be true that the standard of living of the masses is higher, why are so many still obliged to work for starvation wages? If diseases have not increased, and the physical standard has risen, why has the medical profession so enormously increased in number? Why have the hospitals and charitable institutions increased instead of decreasing? and why does the poor's tax form an ever-growing item in expenditure? Why do so many virgin daughters yield up their souls on the altar of lust every year?

The self-contented man who has enjoyed every advantage cannot understand that there are others who desire similar advantages. He often argues with dogmatic decision : The tramp was in the days of our forefathers, and there is no reason why he should not exist in our day. Why trouble ourselves about him ? Prostitution and crime were, are, and ever shall be : stricter and more stringent punitive methods are all that is needed. He forgets that there are none so low or vile but they have some influence on those about them ; and who, even if isolated from society for a time in prison, will return to form part of the life of town or city.

This is the contention of those who deride the argument that the money expended annually in keeping up public institutions for the treatment of diseases or crime might be more intelligently and economically spent in their prevention. Such individuals regard every proposal that does not accord with their own selfishness as not an appeal on behalf of the poor and ignorant so much as a bid for mere self-aggrandisement. Above all things it is necessary to have absolute truth on vital statistics. Hence, we maintain that a humanitarian government would place these questions without delay scientifically before the public. It would establish a test of excellence, and thus it would be able to judge how large a number fell below this standard, and to what degree they are in defect. Beside the mere fact of settling beyond dispute many vital questions, a humanitarian government by establishing such a standard would do more to arouse and educate public opinion on vital subjects than any amount of academic teaching.

There is no greater check to progress than that woman should be obliged to procure her livelihood by trading on her sex, whether by way of marriage for mercenary considerations, or by yet more degrading expedients. The manifold evils to the home and to the individual which follow from such dependence are sapping the vitality of the race.

To ignorance on vital subjects is due a large part of the misery and degradation of individuals. "If I had only known!" is the heart-wail of many a poor soul. A knowledge of consequences is a check upon actions which wreck body and soul alike. How often is not the mother responsible for the dwarfed intellect and rash conduct of a young girl; she neglected to teach and train her child so that she would have been physically and mentally superior. Looking a little deeper into the subject, we find that the mother herself, too, was ignorant, and in turn could reproach her own mother with like neglect. Many miserable mothers have turned and cursed those who have allowed them in ignorance to produce unworthy offspring. Many a man, endowed by nature with superior advantages, has, through ignorance, become a physical wreck. Probably, when he was sent out into the world, his father let him take his chance, as he himself had done.

Could the life-histories be written of those who have been left to take their chance, it would unfold such a tale of horror that the very stoutest heart would quail. A man or woman who walks into a pitfall with no knowledge of its existence is not to blame; but a man or woman who deliberately walks there knowing the danger, rightly suffers the consequences. This is why we have urged that the elementary laws governing our being should be taught to boys and girls alike; and have maintained that such laws, intelligently directed, would stamp out disease and vice, and produce a race superior in every quality to any that have gone before.

How inconsistent are they who discuss openly the breeding of choice animals, but blush at the mere mention that these same laws govern their own breeding! Where constitutional defects are most prominent in the parents, there we find the offspring similarly afflicted. By tracing the pedigrees of sire and dam the characters which the progreny will possess can be foretold. Where it has been desired to eliminate any defects, there all animals who were pre-disposed to such defects have been rejected for breeding purposes. If we desire to stamp out constitutional defects, and to have a superior race of men, the radical remedy is to exclude the unfit from breeding. Hence, how such exclusion can best be accomplished becomes the all important problem.

We do not lose sight of the conditions which make persons unfit when we say that the continuance of the race should be left to the strongest and healthiest; we do not forget the favourable conditions which make individuals the strongest and healthiest.

In breeding superior horses, it is said that good food must be given to the mare; also that impure air, overcrowded stables, or close confinement must be avoided, or the foal will suffer. The mare must be kept in the best possible condition of health and vigour; she must not be overworked during pregnancy if a superior foal is to be produced. What must be the effect of impure air, overcrowded rooms, close confinement, and unscientific feeding on the child of the half-starved human mother?

Not forgetting the influences which produce the unfit, we assert that the methods by which unworthy individuals may be excluded from breeding are:—

1. By individual sense of duty strengthened by educating men and women to the responsibility of becoming parents.

2. By educating public opinion on every possible occasion as to the importance of intelligent breeding, until there is a reaction in public sentiment against the crime of perpetuating infirmities.

3. By the absolute isolation from society of irresponsible or unfit individuals.

If we wish to improve the race, we must do away with false standards of value, such as the possession of money, fictitious patents of nobility, are other inducements which determine modern marriages, but have no real value to breeding *per se*. Who would match horses because their stables were particularly fine, or because they had handsomer and finer clothing than those worn by other horses ? They would be matched for the merits which they were known to possess as horses. Similarly in the human race we must establish standards of individual merit which will be transmitted to offspring.

Legislation forbidding diseased, weak-minded, criminal, or insane persons from going through the legal ceremony of marriage would be useless ; they would marry without the legal ceremony. It is only by proper incentives, high ideals, and a more extensive knowledge of physiology that we shall be able to exterminate diseases and mental defects. It is only by creating a desire to arrive at physical and psychical perfection that the progressive elevation of humanity can be accomplished. One of the conditions in procuring a marriage licence should be that the contracting parties have some knowledge of physiology, while a description of the occupation and physical condition of the parents of the contracting parties should be given. And when a child is registered, a similar description by government physicians should be made, Coming generations would thus be benefited. By this means scientific data could be accumulated which would be of immense value in improving the race.

A visit to any hospital or infirmary must excite a feeling of indignation that a community should be so negligent of its duties to itself. At a vast expense these great buildings are kept up, under the care of able-bodied doctors and nurses. Two, and in some cases three, stand over an idiot child three or four months old; at another cot a nurse and a doctor are watching a child with a cleft palate and hare lip; two-thirds of the children in one ward are, and never can be anything else than, a burden and a tax on the State. It is an exhibition of the grossest ignorance on the part of those who are supposed to be charged with the public weal. Hundreds of thousands of dollars are expended yearly on these monstrosities.

Under a humanitarian government, educated men and women would be employed in every district, parish, and borough, whose duty it would be to find out the physical and mental condition of every man woman, and child, and to report the same to thoroughly trained officials at headquarters, so that the physical, mental, and moral condition of an entire community might be known almost at an hour's notice. An official gazette would regularly publish the reports from headquarters, so that the people of each district might know the condition of their surroundings as perfectly as they now do that of their own isolated homes. The Government would employ trained scientific teachers in all branches of education, whose business it would be to see that none remained ignorant on any subject pertaining to their mental, moral, or physical well-being. Trained nurses would teach every pregnant woman her duty to herself and her unborn child ; and on the birth of each child, its physical condition should be duly reported. By such means there would be implanted in the parents a desire to have as perfect a physical representation of themselves as possible ; and even should there be born a monstrosity of any kind, the public would be in a position to investigate the condition of the parents, and to find out whether, when the demanded a licence to marry, they were both fit subjects for such a purpose. By so doing the people would be made to realise what are the causes to which they owe the ever-growing burden of taxation for the maintenance of the non-effective and useless members of the community.

A real aristocracy of blood would be an object-lesson in heredity. Many can trace their pedigrees five or more generations ; does this give them any title to our regard ? No ; the names showing descent are valueless unless they can show that superiority was transmitted with these titles. The thirteenth Earl of B. or the eighteenth Duke of S. indicate nothing unless the family or life history show that the individual possesses some of the qualities which gave the first the right to be deemed superior to his fellow kind. Better have no pedigree than be the degenerate descendant of a line of kings.

As regards marriage, the legal ceremony has been made the social standard of morality. Plays are passed by the public censor in which the characters are made to talk of selling themselves for money, provided they go through the legal ceremony. Plays are condemned which turn on the bartering herself by a woman without the legal ceremony. To me, it is as immoral that a man should talk of legally selling himself to a woman for her money as that he should discuss doing so without the ceremony; but in this moral age the legal ceremony is all that is required. The young and beautiful mate with the infirm and diseased; if they observe the conventional standard of the legal ceremony, they are good; if they mate without it, they are bad. Among the rich, etiquette has set up so many obstacles and so many deceptions to prevent men and women from acquiring before marriage a correct knowledge of each other's true character, physical and mental, that we need not wonder at the constant activity of the divorce court. Marriages among the well-to-do classes are often social or commercial partnerships, for the purpose of increased importance in the ranks of fashion, or for amalgamating handsome fortunes into one splendid inheritance. Can such marriages be said to be made in heaven? Again, many actions are permitted under the cloak of the legal ceremony which are considered highly immoral without it. Many a woman marries for a home; the man may be indifferent, or even objectionable to her. A lady confessed to me some time since: "I married for a home; I dislike my husband, but I submit to him for my food and clothes; what unfortunate does more? The law throws about me a cloak of respectability, but at heart I am no better than they are."

I by no means wish to do away with all standards of morality with regard to sexual relations. That the interests of society as at present organised render necessary some formal contract in evidence of marriage, is not to be denied ; but my object is to establish a standard which would exalt the purpose for which marriage was instituted. I hold that all marriages are highly immoral which have as a result imperfect and deteriorated offspring. I maintain that when the Church solemnises a marriage of the unfit, it perpetrates a crime. Further, though legally married to a drunkard, an epileptic, or otherwise unfit individual, a woman perpetrates a crime if she continue to have children by such a husband, in spite of the fact that the law sanctions such an action, and the Church says : " Thy desire shall be unto thy husband and he shall rule over thee." So long as the legal ceremony is the only accepted standard of morality in sexual relations, we cannot hope for improvement along scientific lines.

It is often said that if the legal ceremony as the standard of morality be abolished, every man who is tired of his wife will leave her for another, or when she becomes old she will be replaced by some younger and prettier woman. But my view is, that if the financial dependence of woman on man be abolished, those men who could not exercise self-control or who were not actuated by high ideals would be weeded out by the operation of sexual selection. Give freedom of choice to woman by making her procreative function independent of, not subservient to, her daily wants, and then will be bred a better race of men.

Until the question of scientific propagation is divested of its mystery, and discussed from the pulpit, by the school, and in the public Press, very little progress can be made. We cannot have a better race until we are ready to discuss the factors which shall bring this about. It is only by discussion that the power to do harm which secrecy and ignorance confer on this subject will be disarmed. It was an observation of the great Flaxman that " the students in entering the academy, where they studied from the nude figure, seemed to hang up their passions with their hats." Their familiarity with natural beauty led them only to inform their minds, and to purify their taste. Love should receive culture as does music, sculpture, painting, or any other fine art. The world admits that taste has to be educated in these.

As we become better judges of individual excellence, and fictitious values of individuals are superseded by real values, a higher morality will determine the marriage relation.

Woman will then have a more just conception of her maternal responsibility; she will feel that she is accountable for the instruction of her children in all the mysteries of sex, so that none shall go into marriage in ignorance of the laws and uses of the reproductive functions, which now are kept concealed by the false delicacy of ignorance, instead of being made matter for earnest consideration and complete understanding.

For ages it was believed that the earth was flat, and many were the calculations derived from this false premiss. But with the discovery that the earth is round there came a revolution of ideas, and the old philosophy was discarded. Under the old ideas of marriage, held when the first laws of nature were not understood, a system of marriage which exalts the legal ceremony to be the moral standard may suffice; but with the knowledge that we can determine and control the quality and quantity of offspring, that we have it in our power to do almost what the celebrated breeder Bakewell did—who regarded animals as wax in his hands, so that he could produce whatever form he wished—the conviction follows of necessity that the marriage which subjugates woman's maternal function is highly immoral. If we compare the present system of marriage with some low and degraded State, we may conclude that the present system is the best; if we compare the present system with the knowledge of the power of what may be done by intelligent marriage, then it is undoubtedly bad.

Before the application and uses of steam were understood it was a wasted force, but since intelligence has been able to control and direct this hitherto unutilised energy an immense impetus has been given to civilisation. By the intelligent control and direction of this mighty power, and its scientific employment in the service of humanity, even the most ignorant and lowly are benefited. In like manner, an educated will and conscious direction of energy, an endeavour to attain the highest results in ennobling and developing the human race, will do as much for the individual man and woman as force, guided and controlled by the light of science, has been able to accomplish in the external world. An educated will has the power to control the energy too often wasted in the destruction of humanity; the intelligent direction of vital force toward the production of healthy, noble, intelligent citizens will be as remarkable in its results as the application of power to the uses of man. It is the function of educated will to direct energy; if it be directed toward good and noble ends, humanity is by so much enriched; and if it degenerate into mere animalism, humanity is proportionately impoverished.

The uses to which the power generated by steam was at first applied were of the most crude and simple kind ; but to-day this power propels the most complex and elaborate machinery. Among savage tribes, or in a primitive state of society, the reproductive instinct was an elementary function ; as mankind became more highly developed this animal instinct became more complex and elaborate, being associated with qualities emanating from the highest psychic life. Beauty of soul, high principles, noble sentiments, refined and cultured manners, became associated and combined with the simple animal instinct. The more complex the animal instinct, the more it is influenced and restrained by the highest sentiments. The artistic soul has affinities with beautiful landscapes, harmonious sounds, and symmetrical forms. The pleasure or satisfaction derived from beautiful objects does not lie in a single colour or outline, but in the arrangement and blending of colours or forms. So will a man or a woman highly appreciative of honour and elevated thought have dynamic affinities with individuals similarly gifted. The man with music in his soul appreciates the concert room ; uneducated or perverted taste finds satisfaction in the music-hall. He who looks on a woman as merely a being of a different sex is little better than a savage ; he who takes pleasure in the society of a pure and intellectual woman is an ennobled being. The dynamical affinities between highly developed man and woman are physico-psychical ; those between animal man and woman are simply physical.

In a scientific organisation of society, imprudent and ill-assorted marriages will have no place. Educated love alone will therein be the ruling principle. A stigma shall attach to the woman who marries for home or for position. She who sells herself with the

43

legal ceremony shall be deemed as impure, as thoroughly meretricious, as she who sells her person otherwise. The public shall have been educated to so high a standard of honour that public feeling shall be against impure relations and selfish purposes, in whatever form they may appear : purity, virtue, chastity shall be encouraged, and personal honour be strengthened. Marriage shall no longer legally convey the control of the woman's person to her husband. She shall not be subject to an unwelcome touch, and she shall be as much mistress of herself as she was during her maidenhood. Enforced commerce, although cloaked with the legal ceremony, shall be as much a crime as it is now without it, and compulsory child-bearing shall be treated as a double crime committed on the woman. Man shall feel that marriage vows are mutual, and therefore involve mutual faithfulness ; he shall so honour woman that no impure or selfish feeling against her shall ever enter his heart. Marriage shall not then degenerate, as it now in most instances does, into that repulsive condition in which all attraction of the nobler aspirations ceases, and husband and wife become to one another merely physical necessities for the gratification of the lowest animal propensities. Man has only to will to stand in paradise, and he is in it, for it is everywhere, it is in him. But he must cast from him all that may intercept its light from reaching and entering into him ; all grovelling desires must be rent away and thrown into the sacrificial fire kindled on the newly erected altar dedicated to ideal womanhood.

All precautions and preventatives are but sorry expedients, defective and repulsive substitutes for the grand remedial action which must spring from man's

44

purified nature, purified by the contemplation and reabsorption into himself of the ideal woman who, according to the paradisic legend, was taken out of him. When man looks on woman not merely as the means to sexual delights, which in his purer moments he himself feels to be an outrage to the divinity that doth hedge the woman ; when, in fact, he looks on her as a Madonna to be worshipped, then he himself will be ennobled. His body will be exalted into a pantheon, a temple of many divinities, presided over by the two chief goddesses, Sophia and Hygeia—Wisdom and Health—which restore in the human body the balance wherein all the properties and powers of nature work in harmony, and the human frame becomes again what in Genesis is allegorically represented.

In order to have sons and daughters " such as the Dorian mothers bore," we must have mothers such as the Dorian mothers were. This task imposed on women cannot be accomplished without diving into the very depths, and with an awful sense of the responsibility of doing so, and of laying bare its hideous secrets. But as in his physical sphere the chemist out of the most nauseous substances draws forth the most gorgeous colours, and the most delicious scents, since they are there, hidden in the repulsive outward covering, so, in the moral world, evil is evil only because it is misapplied force. Without force there would be no hope of salvation ; but fortunately the force is there, though wrongly directed. To direct this force aright is the task of moral reformers, but the wrenching and the twisting to which in the process they have to subject the social framework, make them appear the enemy of mankind. Hence, no improvement, whether political, social, or industrial, has ever been effected without bringing obloquy, enmity, and even persecution on its first propounders. It is to woman that fate has assigned the difficult but noble task of imparting to mankind the moral momentum which leads to man's artistic and scientific achievements.

When we contemplate the vast mass of misery and destitution in a great city such as this; when we look on the stunted forms and unintelligent faces which defective breeding and defective environment have produced around us, the difficulties of the task appear to be tremendous, if not insuperable. But it is the little leaven that leavens the whole lump. I trust that what I have said may not have fallen altogether on stony ground, but that my words may have awakened in those here present some such longing as that which inspired the great poet, lately lost to England and to the world, to pray, in the pessimism of his declining years, for the advent of—

" Some diviner force to guide us thro' the days I shall not see.
When the schemes and all the systems, kingdoms and republics fall,
Something kindlier, higher, holier—all for each and each for all.
All the full-brain, half-brain races, led by Justice, Love, and Truth;
All the millions one at length, with all the visions of my youth.
All diseases quench'd by Science, no man halt, or deaf, or blind;
Stronger ever born of weaker, lustier body, larger mind."

Reprinted from "THE HUMANITARIAN," August. 1892

Efficacy of Punishments.

What is that part of an individual which is held responsible for his or her actions? Idealists would call this responsible agent the soul; realists would call it the power we have to control our actions, the responsible agent, the higher controlling power of the cerebral cortex. All souls are not alike; we have big-souled and little-souled persons. We are puzzled to know whence the difference arises. Why should one person be endowed with a noble, generous, sensitive soul and another with a contemptible, or mean, or criminal soul, if the soul is independent of the body? We do not punish juvenile delinquents with the same severity or in the same manner as those who have attained maturity. Why do we make this distinction? Is it not because even idealists recognise that the mind or soul has to develop or mature to attain full responsibility, as well as the body to attain its full stature? Anatomy of the brain shows that the cortex of the child is imperfectly developed, and so it is with imbeciles or idiots; they have not the controlling planes fully and perfectly developed Disease, drink, opiates, or any chemicals which attack the higher nervous centres and vitiate the brain, destroy the soul. Disease or accident may so affect a particular group of cells in a growing individual as to arrest their further development altogether ; and this individual on reaching maturity, apparently a man or woman, may still be as irresponsible as the child. The study of the efficiency of punishments, legal or otherwise, is highly instructive. The importance of this study as a preparation to the better comprehension of our fellow creatures cannot be better illustrated than by quoting from Montesquieu's Spirit of Laws: "In the ancient French laws we find the true spirit of monarchy. In cases of pecuniary mulcts, the common people are less severely punished than the nobility. But in criminal cases it is quite the reverse ; the nobleman loses his honour and his voice in court, while the peasant *who has no honour to lose*, undergoes corporal punishment." What were the physiological conditions which made it possible that the peasant had no honour to lose ? Was it not that, being overworked and underfed, his capital of potential energy was at so low an ebb that slight stimuli had no effect on consciousness ? The peasant had not the higher nervous centres sufficiently developed to appreciate the pleasures and pains of the intellect. The corporal punishment administered to the peasant is analogous to the necessarily slight corporal punishment of very young children who have not as yet their higher nervous centres developed which would enable them to reason whether an act is right or wrong.

We punish imbeciles as we punish children. It would be useless to reason with them, because their brains are in a stage of arrested development. It would be ridiculous to talk of the honour or dishonour of an imbecile, because we know the person is incapable of appreciating the one or the other. And not less rediculous is it to speak of the honour or dishonour of moral imbeciles; they have not their higher faculties fully developed. What is remorse? Is it not the reflection which comes after a deed is done? It is the bringing into relation the performance of a deed and its consequences, and the fact that the consequences might have been avoided. Consequences are the check to a repetition of the act. If an act is condoned, and the consequences are slight, in all probability the act will be repeated. Experience not only associates pleasure and pain with an act, but it enables us to say that the consequences of an act will be so and so. It comprises not only the actual state, but the sequences of that state. Consequences which act as a check to the highly organised man may be purely physical; consequences to the less highly developed must be physical pain. Those individuals who have not the power to reflect will not suffer remorse. If reflection is impossible, remorse is impossible.

In corporal punishment we reach the soul or nerve centre which retains the memory of the pain or punishment through the physical pain of the body. Although idealists believe the soul and the body to be distinct, we make the body suffer in corporal punishment for the misdoings of the soul. Russians dread the knout, therefore it is used unsparingly as a means of punishment. This method of punishment is employed because they think it will be remembered by the culprit, and this recollection of former pain will counteract any desire to repeat the offence. It acts on the principle—a burnt child dreads the fire. " In Russia there is a general indifference to death; but the fear of the knout is universal; and in places

10,000 miles from the centre of administration this form of punishment is the prompt, economical, and effective available." The recollection of acute pain associated with an act prevents repetition of the act, and this is the reason corporal punishment is efficacious with certain classes of individuals. The animal learns to avoid the trap where he was once caught, and the food which is injurious, and the like; it learns to avoid that which would cause a repetition of the pain, and it seeks a repetition of that which gives pleasure.

What gives rise to the sensation of pain? If we apply an electric current to one of the peripheral organs and increase the strength of the current, there comes a time when the current can no longer be endured. The current by increasing in intensity has given rise to the sensation of pain. When a few drops of acid are applied to a wound which has some nerves exposed, the intense currents set up by the chemical changes act as irritants to the nerves, which produce sensations of pain. Abnormal conditions affecting the nervous system produce pain. Moderate heat may be pleasurable, moderate cold exhiliarating. Extreme heat gives rise to uneasiness, and to pain, and so, likewise, does extreme cold. The intensity of a stimulus will influence the discharge of a nerve centre, and hence certain co-ordinated movements will result to avoid that which gives pain. Intensity of the stimulus determines whether a nerve centre is going to discharge in a particular manner. Great pain is accompanied or followed by unmeaning movements, not purposeful. And why? Pain seems to be the accumulation of energy in a given organ too great to be distributed at the time.

If an intense current set up can be distributed along different channels, it may not produce pain, or lessen pain; that is, if the surplus energy can be translated into thought or movement it may not produce pain. And that is why a man, or woman, or child, or animal, finds relief in cries, moans, groans, writhings, and bodily contortions. According as energy is distributed to many parts, so the tension in a given area is lowered. It is like letting off steam when the boiler has become too hot, or there is too great pressure in a given area. The potential energy is made actual by the stimulus, and amount of energy made actual and manifested as pain will depend upon the amount of potential energy stored. Intensity of feeling varies in individuals, the normal condition of their nervous systems being different.

The subjective cognition of pleasure and pain will be vivid and intense, according to the individual's capacity of feeling the one or the other. Even with the lash, there arrives a time when there is no more pain felt, because the nervous energy is exhausted. The resultant pain varies with the intensity of the stimulus; and that is why, after having administered a certain number of lashes to one side of the individual, they commence on the other. Benjamin Howard, in an article on Siberian flogging, says: "From a medical standpoint the physical results of flogging have disappointed me, as I have seen it in its various forms in Siberia. In every case the primary and also the secondary shock have been less than I looked for." I account for this by the fact that those individuals who were flogged, were overworked or physically exhausted.

The revival, or remembrance, of a pleasure or pain

will vary in vividness with the health of the organs, and the condition of the central nervous system. Anything which tends to deaden the sensibility of the central nervous system will lessen the pleasure and pain experienced by the individual. Hence the reason why anæsthetics are given to a patient before performing a surgical operation, the drugs deaden the sensibility of the nerves, and there is no cognition of pain, which proves that the appreciation or cognition of pain is dependent on the energy discharged from the nervous centres. A horse which has been resting three days in the stable will dance all over the place at the mere touch of the whip, but a horse tired out with dragging a heavy load, or even ordinarily tired, has to be beaten to be made to go. A horse loses his spirit when the coachman has been cutting down his rations, or physicing him; coachmen will often say: "I must feed the horse less, or you must work him more." I have scarcely a doubt that if a man were put through exercises until he were physically exhausted, and then flogged, the pain would not be intense. It explains why the overworked, underfed, portion of a community is incapable of experiencing the keen or acute pain that the high life individuals experience. It is because, being overworked and underfed, their capital of potential energy is small.

The sensations of pleasure and pain associated with certain actions will vary in intensity with the age of the individual. The keenness of the pleasure or pain is dependent upon the intensity of the stimulus. We recall a particular circumstance which gave us pleasure, even to its details, because the feeling aroused in us at the time was intense, and therefore vividly impressed itself on consciousness. A hundred circumstances have occurred to us since, to give us pleasure, but they are forgotten because the emotion at the time was not of sufficient intensity to leave any great impression, and fades into the many minor impressions of everyday life. The more intense the feeling, the greater is the amount of energy used up. With increasing age the sensations of pleasure and pain are not so keen. Moreover, the little things which caused intense suffering in the child are disregarded in the man or woman. The pleasure of love, and the pains of unrequited love, are most intense at a certain age. The doubts and hopes that rack the mind at one period of our life, we may be indifferent to at another. It is because the intensity of reaction to particular stimuli varies with the condition of the central nervous system, and the organs through which the impression is received.

Disease attacking the auditory centres would cause increasing deafness, and therefore would diminish the pleasure experienced by the performance of beautiful music, or the pain of hearing discords; and total deafness would make an individual oblivious to either. The failure to feel pleasure upon the playing of harmonious, soul-inspiring music is not attributable to the music, but to the conditions of the organs of the individual through which the sensations of sound are received. Degeneration of the optic nerves will diminish the power of seeing beautiful landscapes; and total blindness will make the person insensible to the exquisite pleasures experienced at the sight of beautiful objects, and to intense pain when brought into contact with ugly and repulsive ones, in fact, to all the emotions aroused by the sight of familiar objects. With deafness, the individual becomes insensible to the emotions aroused by familiar music, or the enthusiasm aroused on hearing patriotic airs, the sadness experienced on hearing funeral marches, and the joy and exhiliaration on hearing bright dance music or favourite melodies. A good or bad action, and the intensity of pleasure associated with either, do not depend upon the action itself, but upon the nervous organization of the individual, and the manner in which it has been taught to re-act to these impressions.

The intensity of the sensation of pain is augmented by the attention being directed to it. If we become absorbed in something else, we may forget a pain, or become indifferent to it. As, for instance, the intense pain indured under a terrible excitement, of injuries received in an accident will often go unnoticed, or unheeded, till the excitement is over and then a gash or fracture will be discovered. A patient is often oblivious of pain while undergoing an operation ; but when the excitement is all over, and attention is directed to the wound, the pain is intense. In the heat of battle soldiers have been known to continue fighting, oblivious of pain, though dangerously wounded, where, wrought up to a certain pitch, we become indifferent to pain. Notable instances of this are seen in religious ecstacies, when all the energy is concentrated upon one object. Religious fanatics who mutilate themselves, walk on live coals and the like, and yet seem unconcious of that which, in another, would cause extreme agony. If the extra amount of energy which attention directs to an organ is elsewhere directed, it correspondingly diminishes the energy of the organ affected. But when we become morbid, that state incapacitates us to direct attention elsewhere.

Punishment may take the form of depriving the individual of something to which he or she has been accustomed ; as, when the body has been deprived of food, the pangs of hunger may become so intense as to make all other pains insignificant. Pain conveys no meaning except the contrast which has been drawn between it and pleasure, or a neutral state—the greater the transition, the greater the one or the other. The representation of these two sensation will be vivid, or the reverse, according to the impression made on the individual mind. The deprivation of liberty to the nomad or wild animal, induces much keener suffering than it would to the civilized man or the domesticated animal. The deprivation of drink to the drunkard, causes much keener suffering than to the moderate drinker. The loss of a sum of money to a miser gives rise to a much keener suffering than the man, who is not a miser, is capable of experiencing from the same cause. A plate of food given to a man who has just eaten, will produce a different effect than will food given to a man who has been fasting from want for the previous twenty-four hours. Imprisonment has one effect on the man transferred from a comfortable home to a prison and another on the man who was homeless. When liberty signifies nothing to those deprived of it, when liberty means no home, no chance of getting an honest living, and the being shunned as a pariah, the deprivation of liberty as a mode of punishment is useless, and only puts a tax on the community. The pleasure or pain varies in intensity with the suddenness or greatness of the change. The prison may afford pleasure to the man who has eaten nothing for the previous forty-eight hours. Whereas, with another man, it may have the effect of making him incapable of eating for the following forty-eight hours, that is, until the horror and shame has worn away.

Better conditions of life develop in human beings the faculty of feeling, the pleasures of being praised by our fellow-creatures, and the pain of being blamed. Those who are born and reared in low conditions, cannot experience the grief which high life individuals would experience on losing social caste.

Right or refined conduct has no meaning by itself; it acquires its significance only by being compared with actions which are not good or refined. Those individuals who have been unfortunate enough from childhood to be environed by evil associations, and never have had these associations contrasted with better conditions, have no conception of the moral ought. It is the contrast between the two which teaches the morality. We know night, by having day,—we know right only by having wrong made perceptible to us as a contrast. Those children who are taught to lie, beg, or steal for the benefit of their parents, grow up with the idea that evil is good. And after attaining manhood and womanhood, they carry into practice the training of their youth, and are punished accordingly. The slum child and the child nurtured in refined surroundings, and trained to high conceptions of human conduct, will have quite a different code of morality. The one will be almost incapable of experiencing the exquisite shades of feeling, painful and pleasurable, accompanying a more highly developed nervous organization. It will be interesting to cite instances in proof of this. A lady friend of mine, who had a sewing class in a very poor district, was very much surprised when one of the girls brought a baby to the class. My friend said, turning to the girls, "You know it is forbidden to bring babies to these classes; whose baby is it?" "P—'s" was the answer. My friend with horror and astonishment said, "But P— is not married." My friend was shocked because she had been brought up to look on such a fact with horror. But the girl felt no shame in bringing the baby among a lot of girls whose ages ranked from fifteen to twenty, laughing and joking about the mishap. Her training in morality was different, that was all. Pity being felt for the mother of the chlid, who was only seventeen, occasion was taken to visit her at her lodgings, and the question was asked, "Are you not going to marry the father of your child?" With a shrug she answered, "Indeed not, I hope to do better than that." The baby was simply an incident. How will legal punishment operate taken in connection with perfect callousness of prisoners to shame.

Another case was when, on a round of visits in the country, a woman holding a baby was asked whose baby she was holding, said that it was her daughter's. Some one remarked to her that her daughter was not married. "No," she answered, without the least feeling of shame, "what a pity we couldn't catch her man, and I have to support the brat." There was no comprehension that she had failed to teach her girl what perils beset her path, and guided and reasoned with her. The only suggestion which aroused any emotion in her was that the man could not be made to pay for the keep of the child. We cannot blame her, we must go back of her and blame the system which produces these ignorant mothers. It is only by such examples as these that we can be made to realize the great disparity between individuals, and that there must be something radically wrong in our present social relations. It is important to study the conditions of life which have helped to form the moral character of a given individual, and then we can pass judgment accordingly. It is the contrast made between two opposed states of feeling which gives to each its vivid consciousness. A certain degree of cold has a different effect on a woman who has been out walking and on another woman who has just come out of a warm room. And so it is with all conceptions of human conduct, our actions and judgments are the outcome of individual experience and education.

Dangerous classes will continue to be bred, as long as the conditions of their environment are such as to destroy moral refinement. An individual feels no shame on the performance of particular acts, because no grief, pain, or horror is associated with these acts. The very term dishonoured is only applicable to someone who has been honoured in some way. The individual who has no honour to lose cannot feel the shame of dishonour. We take pleasure in beautiful surroundings, and feel the anguish of being thrust into base, low, ugly surroundings in proportion as the æsthetic faculty is developed in us, and we are capable of appreciating the beautiful. We feel pain at the sight of immoral conduct, and pleasure in seeing or hearing of great and noble deeds, in proportion as the moral faculty is developed in us. All men's faculties of perception are not equal. In trying the effects of sounds in a public hall, in the presence of a large number of persons, it has been proven that the audience were not all affected alike. As the experimenter increased the speed of the vibrations, and the sound became higher, some of the audience remained tranquil, as if deaf, while it gave others the most excruciating agony. Many persons are incapable of hearing very high sounds. Is it not because there no corresponding filaments in the cochlea of the ear to correspond to the vibrations produced? And with the intellect, there some brains so highly organized, sensitized, that every new idea, every æsthetic blending of colours, or noble example, will set some fibres of those millions of the brain quivering in response, and consequently make itself cognisable. The whole universe, as comprehended by us, is only real according to the degree of perception possessed by us. The gamut of the scale of feeling is larger, and more finely strung, in some individuals.

If the æsthetic and moral faculties are not developed in an individual, the individual is not capable of experiencing the pleasures and pains appertaining to æsthetics and morals. And to punish that individual we must affect the physical man. We have to employ coarse and brutal methods of punishment to deal with coarse and brutal persons. If we deplore these methods of punishment, and the large number of the unfit, we must eradicate the conditions which produce them. When human beings are devitalized, their power to resist temptation or pernicious influences is weakened.

The punishments and rewards of religion are ideal, and appeal entirely to the intelligence. The conception of heaven and hell is based on the principles of rewards and punishments—if you act right, if you are righteous, you will be praised, you will be rewarded by going to heaven; if you sin you will be punished by going to hell or purgatory. The hope of reward or fear of punishment inculcated by religion is not real, it is purely ideal. It is not what is but what ought to be. And this thought becomes the actual when the sight of wrong conduct, evil actions, give rise to pain in the human mind, and the opposite gives pleasure. How will those modern biologists who do not admit the Lamarkian theory of use-inheritance account for the fact that many of the ideals of antiquity have become the real of to-day. It is ridiculous to say that those individuals who survived were a happy variation in the direction of morality. It was real in the history of each race to give full vent to animal instincts, and ideal to place any restraint on them. The human in us is the power to check or restrain that which is a natural tendency. And the ideal rewards and punishments, the contrasting of good and evil of religion, have had their influence in developing the human in us.

From the simplest misdemeanor to the most terrible crime there are certain associations attached to each and every one—the pleasurable emotions aroused by good conduct and painful emotions associated with bad conduct. The word sin conveys to the mind several varieties of evil actions and there may be different ways of punishing each, but as the individuals of a particular race are punished, what they are accustomed to associate with the action will be recalled when the particular form of sin is named.

The punishments and rewards of society exert a powerful influence in determining conduct. The influence of public opinion has been very potent in developing character. What people will think, has a great influence on conduct. Public opinion ratifies or augments legal punishments by making the culprit feel all the terrors of social ostracism. Its power is direct and certain, therefore more deterring than all the imaginable possible punishments of the other world combined. It inspires human beings with a desire to do good and noble deeds by rewarding them with praise, love, respect. It trusts a man or woman who is known for honesty and probity, and distrusts them when the contrary, thereby adding to public security. It inspires individuals to emulate anything noble or courageous. It stimulates just and courageous actions. For example, a man or woman who risks his or her life to save that of another, is given a medal for brave conduct. The mere gift of a medal has little value, nor is the value in the medal itself, but public opinion gives the value by honouring the receiver. And so it is with all decorations and rewards of merit, public opinion has the power to ratify and give the value to a public gift. It stimulates the desire to acquire knowledge and position by giving these homage. Kindness is repaid by love, cruelty by hate. A man or woman does not like to be called a fool, a good-for-nothing, selfish, or any other opprobrious epithet ; therefore, he or she avoids doing anything which may call such forth. Thus it places a restraint or checks that which may be a natural tendency.

61

An action is judged in its relations to its antecedents and to its consequents. A man commits a murder in order to rob his victim; he is a criminal, and is regarded with horror. Another man murders a man who has alienated his wife's affections to defend his honour, and he is not regarded with the same degree of horror as the first would be. The latter might not even be called a criminal, the antecedents of the same act " murder," are different; therefore, the judgment of the act varies. Knowledge of the antecedents influence our judgments on a course of conduct adopted by an individual; so will increased knowledge of the influences of environment, and education in forming the character of the individuals, influence our judgments on the course of conduct adopted by those individuals.

Why do we imagine that the penalty of one, two, or three months, or so many years of imprisonment, will reform the criminal and repress crime? Primitive communities did not have the gradation of penalties. Sir Henry Maine tells us " in the almost inconceivable case of disobedience to the award of the village council, the sole punishment, or the sole certain punishment, would be universal disapprobation." In Rome, at one time, we learn that the citizens were so far developed that it was only necessary to point out the right way and the citizens would follow it, or the punishment was sufficient if a man was reported dishonest. We cannot afford to have in any community individuals who have no honour to lose. It is not reformation of the criminal we want. We do not want the formation of criminals.

In archaic communities there were no lawyers, and there was no expensive judicial machinery. It is only when society becomes so intricate and complex as it is to-day that we need these institutions. In primitive societies custom was the lawgiver. The whole community met and decided the merits of the case. It is only as societies become more differentiated that we find written laws, and as these societies advanced revision of the laws has been necessary. All individuals, unfortunately, do not move in the direction of progress. Individuals are apt to revert through disease or abnormal conditions to some primitive type, which would be suited or adapted to some archaic community, and we administer to them our later day code of morals, and punish them by law, which is incongruous.

And this leads us to the question, What right have we to inflict penalties on our fellow creatures? The answer might be—It is the same right which all social groups of animals take upon themselves who associate together for mutual benefit, to excommunicate or kill the one, who by its inability, disability, or refusal to conform to the code of that group, makes itself obnoxious to the rest of the community. And this was the view evidently taken by primitive groups of human beings who associated together for mutual benefit in the early tribal communities and later village communities. Each group had its own social code, and when this code was violated the offender was outlawed. Social groups of to-day act similarly. If a man or woman shocks, very badly, the community he or she is shunned, ostracised, excommunicated from society. Early social groups cut off the offender from all communication with the group, and as Sir Henry Maine has already said, " All communities who can at once place its malefactors outside its bounds have little need of an elaborate criminal jurisprudence." But to-day the offender is still part of that village or town, and even if isolated for a time returns to become part of the community. If excommunicated from one class the influence on another for evil is still present, and if exiled, will have influence on another nation. And as surely will be excommunicated from the societies of to-day, the homeless, the outcasts, exert their influence on the community in which they live.